THEORETICAL MAGNETOFLUIDDYNAMICS

APPLIED MATHEMATICS AND MECHANICS

An International Series of Monographs

EDITORS

Volume 1. K. Oswatitsch: Gas Dynamics, *English version by G. Kuerti* (1956)

Volume 2. G. Birkhoff and E. H. Zarantonello: Jet, Wakes, and Cavities (1957)

Volume 3. R. von Mises: Mathematical Theory of Compressible Fluid Flow, *Revised and completed by Hilda Geiringer and G. S. S. Ludford* (1958)

Volume 4. F. L. Alt: Electronic Digital Computers—Their Use in Science and Engineering (1958)

Volume 5A. Wallace D. Hayes and Ronald F. Probstein: Hypersonic Flow Theory, second edition, Volume I, Inviscid Flows (1966)

Volume 6. L. M. Brekhovskikh: Waves in Layered Media, *Translated from the Russian by D. Lieberman* (1960)

Volume 7. S. Fred Singer (ed.): Torques and Attitude Sensing in Earth Satellites (1964)

Volume 8. Milton Van Dyke: Perturbation Methods in Fluid Mechanics (1964)

Volume 9. Angelo Miele (ed.): Theory of Optimum Aerodynamic Shapes (1965)

Volume 10. Robert Betchov and William O. Criminale, Jr.: Stability of Parallel Flows (1967)

Volume 11. J. M. Burgers: Flow Equations for Composite Gases (1969)

Volume 12. John L. Lumley: Stochastic Tools in Turbulence (1970)

Volume 13. Henri Cabannes: Theoretical Magnetofluiddynamics

Theoretical Magnetofluiddynamics

HENRI CABANNES
FACULTÉ DES SCIENCES
UNIVERSITY OF PARIS
PARIS, FRANCE

Translated by MAURICE HOLT with the assistance of A. A. SFEIR
DIVISION OF AERONAUTICAL SCIENCES
UNIVERSITY OF CALIFORNIA
BERKELEY, CALIFORNIA

1970

ACADEMIC PRESS New York and London

ACADEMIC PRESS, INC.
111 Fifth Avenue, New York, New York 10003

United Kingdom Edition published by
ACADEMIC PRESS, INC. (LONDON) LTD.
Berkeley Square House, London W1X 6BA

LIBRARY OF CONGRESS CATALOG CARD NUMBER: 75-117095

PRINTED IN THE UNITED STATES OF AMERICA

CONTENTS

5. Flow past Bodies

6. Flow past Thin Profiles

TRANSLATOR'S PREFACE

Professor Cabannes' treatise on Theoretical Magnetofluiddynamics was first developed from lecture notes of a course given in the Department of Theoretical Mechanics, University of Paris. The course was part of the third cycle in the Faculty of Science and its level roughly corresponds to that of second- or third-year graduate study in the United States. Previous acquaintance with the theory of compressible fluids will be helpful to the student of this subject, although Professor Cabannes develops the theory from first principles, explaining all steps clearly and in detail.

The book treats magnetofluiddynamics entirely from the macroscopic point of view. Much of it is therefore concerned with the effect of external or self-induced electromagnetic forces on the motion of compressible fluids and the modifications such forces have on classical phenomena in gases, such as shock wave motion and aerodynamic behavior. In most of the book the fluid is regarded as inviscid but viscous effects are discussed at some length in Chapter 5.

The French edition has been used as the required text for the regular course in magnetofluiddynamics given in this division of the university for several years and has been enthusiastically received by both faculty and students. It is hoped that this translation, which is of the second, expanded French edition (published in 1969), will make the book available to a wider range of students.

Professor Cabannes has been very cooperative at all stages in the preparation of the translation and it has been a pleasure to collaborate with him. Dr. Sfeir was most helpful in the preliminary stages of the translation and his assistance certainly speeded up the completion of this project. Finally, I am especially indebted to Mrs. Arlene Martin, who was responsible for most of the typing of the manuscript.

PREFACE

Magnetofluiddynamics is concerned with the study of the interaction between magnetic fields and fluid conductors of electricity. This discipline serves to unite classical fluid mechanics with electromagnetism of media in motion; therefore, it is equally of interest to students of electrodynamics and to those of aerodynamics. The former always think of speeds which are of the order of the speed of light while the latter confine their attention to speeds which are of the order of the speed of sound.

From a historical point of view it seems that the first attempt to study the problem of magnetofluiddynamics is due to Faraday. In the nineteenth century Faraday proposed to measure the difference of electrical potential which exists in the water of the Thames when crossing from one bank to the other; he wished to deduce from this the speed of flow. The question concerned the problem of interaction between the earth's magnetic field and the motion of the Thames; depending on the tide, this river flows in either a forward or reverse direction, and the presence of salt water gives it a certain electric conductivity. Since these effects are scarcely discernible, no satisfactory measurements are possible. Faraday's idea was taken up again in 1937 by Hartmann under good conditions; Hartmann carried out experiments which demonstrated the influence of a very intense magnetic field on the motion of mercury. On the theoretical side, the first problem in magnetofluiddynamics was treated by Alfvén in 1940. In the case of steady motion of a liquid having negligible electric resistivity, Alfvén showed that, in the presence of a magnetic field, new waves could be propagated, waves which appeared neither in the mechanics of fluids nor in electromagnetism. Over the past few years the need to study motion in an electromagnetic field of strongly ionized gases at high temperature has become increasingly important, due to the numerous problems arising in physics and engineering (astrophysics, aerodynamics at high speeds, control of thermal nuclear reactions.) For these reasons magnetofluiddynamics is now one of the most active branches of mechanics and physics. The increasing interest which research workers of different disciplines showed in magnetofluiddynamics has led us to present a course in magnetofluiddynamics, as part of the program of the third cycle in theoretical mechanics at the Faculty of Science in the University of Paris.

Obviously, we could not hope to present all the results about magnetofluiddynamics or its applications. When establishing the plan of the course and organizing the different chapters, we were bound to present results relating to new laws and to new qualitative effects which appeared in the course of interaction of a conducting medium with an electromagnetic field. Particular importance has been given to the study of exact solutions of the equations of magnetofluiddynamics. In the first chapter the equations of magnetofluiddynamics are established in detail, both for motion with shock waves and for continuous motion. The two following chapters are devoted to the study of discontinuities: weak discontinuities or discontinuities in the derivatives of the characteristic flow quantities in Chapter 2, strong discontinuities or discontinuities in the characteristic flow quantities themselves in Chapter 3. One-dimensional motion, motion in ducts, and structure of shock waves are considered in Chapter 4, while motion past obstacles is discussed in the last two chapters. We have not treated problems concerning technical applications or results of experimental studies. Moreover, the problems which interest research workers studying the control of thermonuclear reactions, such as magnetohydrostatics and stability, have not been considered. These problems are well understood at the present time and are taught by our colleagues in physics.

The section entitled References lists the papers and articles which have been used directly in developing the various chapters of the course.

In the second chapter, Section 2.2 was based on an article of Bohachevsky (1962), except for Section 2.2.2, taken from the work of Jeffrey and Taniuti (1964). In the third chapter the results related to shock surfaces and the solution of the shock equations, Sections 3.1.2 and 3.1.5, are due essentially to Anderson (1963); the results concerning detonation and deflagration waves were taken from the thesis of Soubbaramayer (1967). Section 3.2, devoted to shock stability, sets out a paper by Akhiezer *et al.* (1959). The solution of the piston problem, Sections 3.3.2 and 3.3.3, is due to Lyubimov and Kulikovskii (1959).

In the fourth chapter, Section 4.2.2 reproduces a work by Gilbarg (1951), and the results of Sections 4.2.4 and 4.2.5 are again taken from the thesis of Soubbaramayer. Section 4.3, flow in nozzles, was based on an article by Meyer (1952) for material in Section 4.3.1, and on Rubin's thesis (1963) for material in Section 4.3.2. The theory of the electromagnetic flowmeter is due to Fabri and Siestrunck (1960).

In the fifth chapter, the first section deals with work by Imai (1960) in Section 5.1.1, by Grad (1960) in Section 5.1.2, and by Hasimoto (1960) in Section 5.1.3. In the second section 5.2.2 is a review of an article by Johnson (1966), while the results given in Section 5.2.3 were obtained by Wilson (1964). The third section was derived from the works of Greenspan and Carrier (1959, 1960). The fourth section, the problem of the magnetized sphere, covers the work of Ludford and Murray (1960), completed by Bois (1970); the results of the fifth section are due to Mme. Levy (1967, 1970).

In the sixth and final chapter, each of the three sections is based on a review of the works of the following authors: McCune and Resler (1960), Sears and Resler (1958), Stewartson (1960).

ACKNOWLEDGMENTS

This English edition corresponds to the course which I taught in 1968 in the University of California at Berkeley and which I am currently teaching at the University of Paris.

I want to thank my colleagues and friends, Professor Maurice Holt and Dr. A. A. Sfeir, for the devotion and competence with which they have assumed the heavy burden of translating the French text into English and thereby making possible the English edition. The reader will be able to admire the care with which the publishers, Academic Press, prepared this edition. I express my warm thanks to them.

CHAPTER 1

GENERAL EQUATIONS

For a continuous distribution of matter, it is always possible to think of an ensemble consisting of the same material elements and to apply to this ensemble the laws of classical mechanics. The equations of motion obtained in this way are integral equations, which we derive in Section 1.1. When the unknown functions are continuous and differentiable, we deduce from the integral equations the partial differential equations stated in Section 1.2. These are first written in vector form, and then in scalar form, referred to a system of arbitrary curvilinear coordinates. When the unknown functions are discontinuous, we deduce from the integral equations algebraic relations satisfied by the discontinuities; these relations, called the shock equations, are established in Section 1.3.

1.1. Integral Form of the Equations of Motion

1.1.1. Definitions

The general study of the mechanics of fluids and of magneto-fluiddynamics can be approached from two different standpoints. In the first approach, called microscopic, it is supposed that the fluid is made up of a discrete ensemble of material particles, the motion of which is determined by the laws of classical mechanics; the fact that the number of particles is large always justifies the application of the theories of statistical mechanics. The problem consists of predicting the probable behavior of the system formed by the particles, starting from an incomplete description of its state at a given instant.

In the second method, called macroscopic, the fluid is represented as a continuous distribution of matter. Only the second method will be used in this book. Magnetofluiddynamics is thus a branch of the mechanics of continuous media. A fluid is distinguished from other continuous media by the property that, in all states of equilibrium, the tensor of the internal stresses is spherical; this property, put in analytical form, and introduced into the equations of mechanics of continuous media, defines fluids.

To represent the motion of a fluid, we use continuous transformations of a Euclidean space in three dimensions onto itself:

$$\mathbf{x} = \phi(\mathbf{X}, t)$$

The transformations adopted depend on a parameter t, which is identified as the time; $\mathbf{x}$ denotes the position at the instant t of a mass element which at the instant $t = 0$ occupied the position $\mathbf{X}$. All quantities F characterizing the state of the fluid can be expressed either as a function of the variables $\mathbf{X}, t$ (material variables, or Lagrangian variables), or as a function of the variables $\mathbf{x}, t$ (spatial variables, or Eulerian variables). The symbols

$$\frac{\partial F}{\partial t} = \frac{\partial F(\mathbf{x}, t)}{\partial t}, \qquad \frac{dF}{dt} = \frac{\partial F(\mathbf{X}, t)}{\partial t}$$

will be used to denote the two possible derivatives of the function F with respect to time.

At each instant t and at each point $\mathbf{x}$ in the space occupied by the fluid, we define the speed $\mathbf{V} = d\mathbf{x}/dt$ and the density ρ. These quantities, as well as all those which will be defined later, will be expressed as a function of the spatial variables $\mathbf{V}(\mathbf{x}, t)$, $\rho(\mathbf{x}, t)$.

When the density is constant, the fluid is called incompressible; when the density is variable, the fluid is called compressible. The internal mechanical forces which act in a continuous medium are determined by a field of tensors $\mathbf{T}$, called stress tensors. The external mechanical forces are given quantities, supposed proportional to the mass; a force $\rho\mathbf{F}\, d\tau$ acts on an element of volume $d\tau$. In the majority of normal cases, these external forces reduce simply to forces due to gravity and can be neglected when the density of the fluid is small. The specific internal energy of the fluid will be denoted by ϵ_i: the various thermodynamic quantities (absolute temperature, specific entropy,...) will be defined from the function ϵ_i .

In magnetofluiddynamics, we must add to the preceding quantities the fundamental electromagnetic quantities, and we shall introduce the following: q, specific charge (electric charge per unit mass of fluid); $\mathbf{J}$, density of electric current; $\mathbf{B}$, magnetic induction; $\mathbf{E}$, electric field. To write the dimensional equations for the newly introduced quantities, it suffices to add to the fundamental quantities introduced in mechanics just one new quantity, for example, electric charge of dimension [Q]; we thus obtain

$$[q] = [Q\,\mathrm{L}^{-3}]$$
$$[J] = [Q\,\mathrm{L}^{-2}\,\mathrm{T}^{-1}]$$
$$[B] = [Q^{-1}\,\mathrm{T}^{-1}\,M]$$
$$[E] = [Q^{-1}\,\mathrm{L}\,\mathrm{T}^{-2}\,M]$$

Finally, we shall make a hypothesis which expresses the principle of impenetrability of matter—namely, that two elements of mass cannot occupy the same position at the same instant.

The physical laws on which the equations of magnetofluiddynamics are based relate to infinite elements of matter; they are expressed in terms of integral equations. These laws must satisfy Galilean invariance, which is the rule of classical mechanics.

1.1.2. Electromagnetic Equations

Electromagnetic phenomena are governed by physical laws, namely, conservation of electric charge and Faraday's law. The first is expressed by the equation

$$\frac{d}{dt}\iiint_{(v)} q\,d\tau = \iiint_{(v)} \frac{\partial q}{\partial t}\,d\tau + \iint_{(s)} q\,\mathbf{n}\cdot\mathbf{V}_{\mathrm{e}}\,ds = 0$$

Since the product $q\mathbf{V}_{\mathrm{e}}$ is equal to the density of electric current, $\mathbf{J} \equiv q\mathbf{V}_{\mathrm{e}}$, we obtain

$$\iiint_{(v)} \frac{\partial q}{\partial t}\,d\tau + \iint_{(s)} \mathbf{J}\cdot\mathbf{n}\,ds = 0 \qquad (1.1.2\text{-}1)$$

where (v) denotes a closed volume, the boundary of which is (s). $\mathbf{n}$ represents a unit vector normal to the surface (s) and directed toward the outside of the volume (v).

Faraday's law is expressed by

$$\frac{d}{dt}\iint\limits_{(\Sigma)} \mathbf{B}\cdot\mathbf{n}\,d\sigma + \int\limits_{(c)} \mathbf{E}\cdot d\mathbf{l} = 0 \tag{1.1.2-2}$$

(Σ) denotes a surface which is not closed, passing through the contour (c); the direction of the length element $d\mathbf{l}$ along the contour (c) and the unit vector $\mathbf{n}$ normal to Σ are related by the direction rule of Stokes' theorem. Equation (1.1.2-1) shows that the charge q is the divergence of a vector field

$$q = \nabla\cdot\mathbf{D}$$

The vector $\mathbf{D}$ so defined is called the electric induction. The flux of the vector $\mathbf{J} + \partial\mathbf{D}/\partial t$ across the whole closed surface is zero; it follows that this vector is the curl of a certain vector $\mathbf{H}$. If (s) denotes a closed surface limited by a contour (c) which remains fixed, the flux of the vector $\mathbf{J} + \partial\mathbf{D}/\partial t$ remains constant; this flux is equal to the circulation of the vector $\mathbf{H}$ along the contour (c),

$$\iint\limits_{(s)} \mathbf{J}\cdot\mathbf{n}\,ds + \iint\limits_{(s)} \frac{\partial\mathbf{D}}{\partial t}\cdot\mathbf{n}\,ds = \int\limits_{(c)} \mathbf{H}\cdot d\mathbf{l} \tag{1.1.2-3}$$

The above formula is a statement of Ampère's theorem; the vector $\mathbf{H}$ defined by this formula is called the magnetic field.

The four quantities $\mathbf{E}$, $\mathbf{B}$, $\mathbf{D}$, $\mathbf{H}$ which have been introduced are not independent, but are connected by relations dependent on the medium. To write such characteristic relations, it is strictly necessary to place oneself in a local region bound with the fluid and locally carried with it. However, when the speeds are small with respect to the speed of light, which is the case in magnetofluiddynamics, we can write such relations in a fixed region. We shall normally limit ourselves to fluids which satisfy the following characteristic relations:

$$\mathbf{D} = \epsilon(\rho, T)\mathbf{E}, \qquad \mathbf{B} = \mu(\rho, T)\mathbf{H} \tag{1.1.2-4}$$

The scalars ϵ (dielectric constant) and μ (magnetic permeability) are known functions of the density ρ and the temperature T.

To determine the effect of an electromagnetic field on the mechanical system, it is necessary to add to the preceding laws the laws which govern this behavior. We shall postulate the following two laws:

(1) The magnetic field exerts on the medium external forces defined by the volume density,

$$\mathbf{\Phi} = q\mathbf{E} + \mathbf{J} \times \mathbf{B} \tag{1.1.2-5}$$

(2) The magnetic field supplies to the medium a power defined by the volume density,

$$\epsilon_{\mathrm{m}} = \mathbf{E} \cdot \mathbf{J} \tag{1.1.2-6}$$

To these two laws it is convenient to add Ohm's law, which connects the current density $\mathbf{J}$ to the electromagnetic field and the speed of the fluid,

$$\mathbf{J} = \sigma\{\mathbf{E} + \mathbf{V} \times \mathbf{B}\} \tag{1.1.2-7}$$

The scalar σ is the coefficient of electrical conductivity; its inverse σ^{-1} is the coefficient of electrical resistivity.

1.1.3. Equations of Dynamics

The equations of dynamics express the momentum theorem and the principles of conservation of mass and of energy. To write these equations we consider a volume (v), always formed by the same elements of matter, and we denote by (s) the surface which bounds the volume (v). To obtain the equations of magnetofluiddynamics, it suffices to write down the classical equations of fluid mechanics and to add the terms given by the relations (1.1.2-5) and (1.1.2-6). We are led to the following three equations:

$$\frac{d}{dt}\iiint_{(v)} \rho\mathbf{V}\, d\tau = \iiint_{(v)} (\rho\mathbf{F} + \mathbf{\Phi})\, d\tau + \iint_{(s)} \mathbf{T}\cdot\mathbf{n}\, ds \tag{1.1.3-1}$$

$$\frac{d}{dt}\iiint_{(v)} \rho\, d\tau = 0 \tag{1.1.3-2}$$

$$\frac{d}{dt}\iiint_{(v)} \rho\left(\epsilon_{\mathrm{i}} + \frac{V^2}{2}\right) d\tau = \iiint_{(v)} (\rho\mathbf{F}\cdot\mathbf{V} + \epsilon_{\mathrm{m}})\, d\tau + \iint_{(s)} (\mathbf{T}\cdot\mathbf{n})\cdot\mathbf{V}\, ds - \iint_{(s)} \boldsymbol{\theta}\cdot\mathbf{n}\, ds \tag{1.1.3-3}$$

The vector $\boldsymbol{\theta}$ denotes the heat flux measured in mechanical units.

1.2. The Equations of Motion

1.2.1. Preliminary Formulas

The equations of magnetofluiddynamics written in the preceding Section are integral equations. When the unknown functions are continuous and differentiable, it is possible to replace the integral equations by partial differential equations. To do this, we must use some theorems of analysis, which we shall recall.

To start with, we shall point out five formulas which transform volume integrals into surface integrals. The first are applied to a fixed or moving volume (v), and the second to the boundary (s) of the volume (v); $\mathbf{n}$ denotes the unit vector directed along the outward normal to the surface (s):

$$\iint_{(s)} (\mathbf{n} \times \mathbf{A})\, ds = \iiint_{(v)} \nabla \times \mathbf{A}\, d\tau \tag{1.2.1-1}$$

$$\iint_{(s)} \mathbf{n}\, \varphi\, ds = \iiint_{(v)} \nabla\varphi\, d\tau, \qquad \iint_{(s)} (\mathbf{n}, \mathbf{A})\, ds = \iiint_{(v)} \nabla\mathbf{A}\, d\tau \tag{1.2.1-2}$$

$$\iint_{(s)} \mathbf{n} \cdot \mathbf{A}\, ds = \iiint_{(v)} \nabla \cdot \mathbf{A}\, d\tau, \qquad \iint_{(s)} \mathbf{n} \cdot \mathbf{T}\, ds = \iiint_{(v)} \nabla \cdot \mathbf{T}\, d\tau \tag{1.2.1-3}$$

We next note the formulas for differentiating triple integrals over a variable volume under the integral sign. The volumes (v) which we shall consider in the application of these formulas are always formed by the same fluid particles and we suppose that these particles do not mix; with this hypothesis, which amounts to neglecting the phenomenon of diffusion, the same molecules are always to be found on the surface (s). The speed of displacement of this surface is then equal to the scalar product of the speed $\mathbf{V}$ of the fluid and the unit normal vector $\mathbf{n}$. We thus have the following identities:

$$\begin{aligned} \frac{d}{dt} \iiint_{(v)} \varphi\, d\tau &= \iiint_{(v)} \frac{\partial\varphi}{\partial t}\, d\tau + \iint_{(s)} \varphi\mathbf{V} \cdot \mathbf{n}\, ds \\ \frac{d}{dt} \iiint_{(v)} \mathbf{A}\, d\tau &= \iiint_{(v)} \frac{\partial\mathbf{A}}{\partial t}\, d\tau + \iint_{(s)} \mathbf{A}(\mathbf{V} \cdot \mathbf{n})\, ds \end{aligned} \tag{1.2.1-4}$$

We recall finally the definition of the different tensor notations used in the preceding formulas:

$$\mathbf{T} = \begin{vmatrix} T_{xx} & T_{xy} & T_{xz} \\ T_{yx} & T_{yy} & T_{yz} \\ T_{zx} & T_{zy} & T_{zz} \end{vmatrix}$$

We denote by $\overline{\mathbf{T}}$ the tensor, the components of which are deduced from those of tensor $\mathbf{T}$ by symmetry with respect to the principal diagonal:

$$\begin{aligned} (\nabla \cdot \mathbf{T})_x &= \frac{\partial T_{xx}}{\partial x} + \frac{\partial T_{xy}}{\partial y} + \frac{\partial T_{xz}}{\partial z} \\ (\nabla \cdot \mathbf{T})_y &= \frac{\partial T_{yx}}{\partial x} + \frac{\partial T_{yy}}{\partial y} + \frac{\partial T_{yz}}{\partial z} \\ (\nabla \cdot \mathbf{T})_z &= \frac{\partial T_{zx}}{\partial x} + \frac{\partial T_{zy}}{\partial y} + \frac{\partial T_{zz}}{\partial z} \end{aligned} \tag{1.2.1-5}$$

By $\mathbf{B} = \mathbf{T} \cdot \mathbf{A}$, we mean

$$\begin{aligned} B_x &= T_{xx}A_x + T_{xy}A_y + T_{xz}A_z \\ B_y &= T_{yx}A_x + T_{yy}A_y + T_{yz}A_z \\ B_z &= T_{zx}A_x + T_{zy}A_y + T_{zz}A_z \end{aligned} \tag{1.2.1-6}$$

By definition, $\mathbf{A} \cdot \mathbf{T} = \overline{\mathbf{T}} \cdot \mathbf{A}$.

If the tensor $\mathbf{T}$ is symmetric, we have

$$\overline{\mathbf{T}} = \mathbf{T}$$
$$\mathbf{A} \cdot \mathbf{T} = \mathbf{T} \cdot \mathbf{A}$$
$$(\mathbf{T} \cdot \mathbf{A}) \cdot \mathbf{B} = (\mathbf{T} \cdot \mathbf{B}) \cdot \mathbf{A}$$

The term $\mathbf{S} : \mathbf{T}$ represent the following scalar:

$$\mathbf{S} : \mathbf{T} = \sum_{\alpha} \sum_{\beta} S_{\alpha\beta} T_{\beta\alpha}$$

Finally, we point out the following definitions:

$$(\mathbf{A}, \mathbf{B}) = \begin{vmatrix} A_xB_x & A_xB_y & A_xB_z \\ A_yB_x & A_yB_y & A_yB_z \\ A_zB_x & A_zB_y & A_zB_z \end{vmatrix} \tag{1.2.1-7}$$

$$\nabla\mathbf{A} = \begin{vmatrix} \partial A_x/\partial x & \partial A_y/\partial x & \partial A_z/\partial x \\ \partial A_x/\partial y & \partial A_y/\partial y & \partial A_z/\partial y \\ \partial A_x/\partial z & \partial A_y/\partial z & \partial A_z/\partial z \end{vmatrix} \tag{1.2.1-8}$$

1.2.2. Constitutive Relations

The equations which we obtained in Section 1.1, electromagnetic equations and dynamical equations, are insufficient in number to permit the study of motion of the fluid. Just as we were obliged to add electromagnetic equations characteristic of the medium, it is necessary to add analogous relations of a mechanical and thermodynamic nature. The relations of mechanical type are called the constitutive relations of the medium; they must satisfy the conditions stated at the beginning of this chapter which enable us to distinguish fluids from other continuous media.

To write the constitutive relations, we introduce the tensor $\mathbf{D}$, equal to half the sum of the gradient of the fluid speed and the conjugate tensor

$$\mathbf{D} = \tfrac{1}{2}\{\nabla\mathbf{V} + \overline{\nabla\mathbf{V}}\} \qquad (1.2.2\text{-}1)$$

We resolve the tensor $\mathbf{D}$ into the sum of a spherical tensor (tensor proportional to the unit tensor $\mathbf{U}$) and tensor of zero trace $\mathbf{D}_1$. In the same way, we resolve the stress tensor into a spherical tensor and a tensor of zero trace $\mathbf{T}_1$ called the viscous stress tensor:

$$\mathbf{D} = \tfrac{1}{3}(\nabla \cdot \mathbf{V})\,\mathbf{U} + \mathbf{D}_1$$

$$\mathbf{T} = \alpha\mathbf{U} + \mathbf{T}_1$$

The tensor $\mathbf{D}$ is called the rate-of-strain tensor. The fluids which we consider in this book are called Newtonian and are defined by the following laws of behavior:

The spherical part of the stress tensor is a linear function of the spherical part of the rate-of-strain tensor. The nonspherical part (with zero trace) of the stress tensor is proportional to that of the rate-of-strain tensor. These are represented by

$$\alpha = -p + \tfrac{1}{3}(3\lambda_1 + 2\mu_1)\nabla \cdot \mathbf{V}$$

$$\mathbf{T}_1 = 2\mu_1\mathbf{D}_1$$

where

$$\mathbf{T} = (-p + \lambda_1\,\nabla \cdot \mathbf{V})\mathbf{U} + 2\mu_1\mathbf{D} \qquad (1\ 2.2\text{-}2)$$

The scalars λ_1 and μ_1 are called coefficients of viscosity; they are functions, supposed known, of the absolute temperature. The coefficient μ_1 is positive or zero. When $\lambda_1 = \mu_1 = 0$, the fluid is called perfect, and when $\lambda_1^2 + \mu_1^2 \neq 0$, the fluid is said to be viscous. The scalar p, which is defined as the coefficient of the terms of degree zero in the linear relation connecting the spherical parts of the two tensors, is called the pressure of the fluid.

To the constitutive relations, we are going to add relations of thermodynamic type which also determine the characteristics of the media. The first law shows that the specific internal energy ϵ_i is wholly determined when one knows the specific volume $\tau = \rho^{-1}$ and the pressure p, and that the function $\epsilon_i(\tau, p)$ is such that the expression $d\epsilon_i + p\, d\tau$ admits an integrating factor. According to this, the first law of thermodynamics, there therefore exist two functions T and s such that

$$(1/T)\{d\epsilon_i + p\, d\tau\} = ds \qquad (1.2.2\text{-}3)$$

The quantities T and s thus defined in a unique fashion (neglecting the additive constants in s) are called the absolute temperature and the specific entropy, respectively. When the specific internal energy is expressed as a function of s and of τ, we have

$$\partial\epsilon_i/\partial s = T, \qquad \partial\epsilon_i/\partial\tau = -p \qquad (1.2.2\text{-}4)$$

By eliminating the specific entropy between these two relations, we obtain an equation

$$p = p(\rho, T)$$

called the equation of state of the fluid. Knowledge of the equation of state permits us to write the partial differential equation

$$\partial\epsilon_i/\partial\tau + p(\tau^{-1}, \partial\epsilon_i/\partial s) = 0$$

satisfied by the function $\epsilon_i(s, \tau)$. In practice, the function $\epsilon_i(s, \tau)$ is considered as given, determined by experiment. The pressure can then be expressed in finite terms with the aid of the quantities s and ρ, which will be considered as independent thermodynamic variables.

The propagation of heat in the fluid is connected with the distribution of temperature. We will suppose that the flux of the quantity of heat $\boldsymbol{\theta}$ is proportional to the gradient of the temperature,

$$\boldsymbol{\theta} = -\lambda\, \nabla T \qquad (1.2.2\text{-}5)$$

The scalar λ, called the coefficient of thermal conductivity, is a function, supposed known, of the temperature T; this coefficient is positive or zero.

1.2.3. General Equations of Continuous Motion

Starting with the integral equations of Section 1.1 and the analytical formulas recalled at the beginning of this section, we obtain the partial differential equations which govern continuous motion in magneto-fluiddynamics, in the following form:

$$\frac{\partial \mathbf{V}}{\partial t} + \nabla \cdot (\rho \mathbf{V}, \mathbf{V}) = \rho \mathbf{F} + \boldsymbol{\Phi} + \nabla \cdot \mathbf{T}$$

$$\frac{\partial \rho}{\partial t} + \nabla \cdot \rho \mathbf{V} = 0 \qquad (1.2.3\text{-}1)$$

$$\frac{\partial}{\partial t}\left\{\rho\left(\epsilon_{\mathrm{i}} + \frac{V^2}{2}\right)\right\} + \nabla \cdot \left\{\rho \mathbf{V}\left(\epsilon_{\mathrm{i}} + \frac{V^2}{2}\right)\right\} = \rho \mathbf{F} \cdot \mathbf{V} + \epsilon_{\mathrm{m}} + \nabla \cdot \mathbf{T} \cdot \mathbf{V} - \nabla \cdot \boldsymbol{\theta}$$

$$q = \nabla \cdot \mathbf{D}$$

Should be ∇xH

$$\mathbf{H} = \mathbf{J} + \frac{\partial \mathbf{D}}{\partial t} \qquad (1.2.3\text{-}2)$$

$$\frac{\partial \mathbf{B}}{\partial t} + \nabla \times \mathbf{E} = 0$$

These twelve scalar equations contain 21 unknown scalar quantities,

$$\mathbf{T}(\text{six}),\ \mathbf{H}, \mathbf{E}, \mathbf{V}, \boldsymbol{\theta}\ (3 \times 4 = 12),\ \rho,\ \epsilon_i,\ q(\text{three})$$

$$\mathbf{J} = \sigma\{\mathbf{E} + \mathbf{V} \times \mathbf{B}\}, \qquad \boldsymbol{\Phi} = q\mathbf{E} + \mathbf{J} \times \mathbf{B}, \qquad \epsilon_{\mathrm{m}} = \mathbf{J} \cdot \mathbf{E}$$

$$\mathbf{D} = \epsilon(\rho, T)\mathbf{E}, \qquad \mathbf{B} = \mu(\rho, T)\mathbf{H}$$

Quantities connected with the unknowns by finite relations are not considered as unknowns. To the twelve (scalar) equations (1.2.3-1) and (1.2.3-2), it is necessary to add the nine scalar equations which express the constitutive relations. If we take the specific entropy s as an unknown in place of the specific internal energy ϵ_{i}, a known function of s and $\tau = \rho^{-1}$, we have

$$T = \partial \epsilon_{\mathrm{i}} / \partial s, \qquad p = -\partial \epsilon_{\mathrm{i}} / \partial \tau$$

$$\mathbf{T} = \{-p + \lambda_1(T)\, \nabla \cdot \mathbf{V}\}\mathbf{U} + 2\mu_1(T)\mathbf{D}$$

$$\boldsymbol{\theta} = -\lambda(T)\, \nabla T$$

We obtain finally a system of 21 equations and 21 unknowns. When the medium is neutral from the electrical point of view, the charge density q is zero, and the number of equations reduces to 20; we shall often suppose that this is the case. It is possible to simplify the first and the third equations of (1.2.3-1).

Taking account of the equation of conservation of mass and the identities

$$\nabla \cdot (\rho \mathbf{V}, \mathbf{V}) \equiv \rho \mathbf{V} \cdot \nabla \mathbf{V} + \mathbf{V}\, \nabla \cdot (\rho \mathbf{V})$$

$$\mathbf{V} \cdot \nabla \mathbf{V} \equiv (\nabla \times \mathbf{V}) \times \mathbf{V} + \nabla (V^2/2)$$

the first term in the first of (1.2.3-1) can be written

$$\rho \left\{ \frac{\partial \mathbf{V}}{\partial t} + (\nabla \times \mathbf{V}) \times \mathbf{V} + \nabla \left(\frac{V^2}{2} \right) \right\}$$

This is simplified by using the equation of conservation of energy. The first term can be written

$$\left(\frac{\partial \epsilon_{\mathrm{i}}}{\partial t} + \mathbf{V} \cdot \nabla \epsilon_{\mathrm{i}} \right) + \rho \mathbf{V} \cdot \left\{ \frac{\partial \mathbf{V}}{\partial t} + \nabla \left(\frac{V^2}{2} \right) \right\}$$

$$= \rho \left(\frac{\partial \epsilon_{\mathrm{i}}}{\partial t} + \mathbf{V} \cdot \nabla \epsilon_{\mathrm{i}} \right) + \mathbf{V} \cdot \{ \rho \mathbf{F} + \mathbf{\Phi} + \nabla \cdot \mathbf{T} \}$$

If we use the identity

$$\mathbf{V} \cdot (\nabla \cdot \mathbf{T}) - \nabla \cdot (\mathbf{T} \cdot \mathbf{V}) = -\mathbf{T} : \nabla \mathbf{V}$$

we write the equation of conservation of energy in the form

$$\rho \left\{ \frac{\partial \epsilon_{\mathrm{i}}}{\partial t} + \mathbf{V} \cdot \nabla \epsilon_{\mathrm{i}} \right\} = \mathbf{T} : \nabla \mathbf{V} - \nabla \cdot \boldsymbol{\theta} + \epsilon_{\mathrm{m}} - \mathbf{V} \cdot \mathbf{\Phi}$$

We have

$$\epsilon_{\mathrm{m}} - \mathbf{V} \cdot \boldsymbol{\phi} = \mathbf{J} \cdot \mathbf{E} - \mathbf{V} \cdot (\mathbf{J} \times \mathbf{B}) = \sigma^{-1} J^2$$

$$d\epsilon_{\mathrm{i}} = T\, ds + p\, d\rho/\rho^2$$

from which we find that

$$\rho T \left\{ \frac{\partial s}{\partial t} + \mathbf{V} \cdot \nabla s \right\} = 2\mu_1 \mathbf{D} : \nabla \mathbf{V} + \lambda_1 (\nabla \cdot \mathbf{V})^2 - \nabla \cdot \boldsymbol{\theta} + \sigma^{-1} J^2 \quad (1.2.3\text{-}3)$$

The equations of magnetofluiddynamics are thus reduced to the following:

$$\rho \left\{ \frac{\partial \mathbf{V}}{\partial t} + (\nabla \times \mathbf{V}) \times \mathbf{V} + \nabla \left(\frac{V^2}{2} \right) \right\} = \rho \mathbf{F} + \mathbf{J} \times \mathbf{B} + \nabla \cdot \mathbf{T}$$

$$\frac{\partial \rho}{\partial t} + \nabla \cdot \rho \mathbf{V} = 0$$

$$\rho T \left\{ \frac{\partial s}{\partial t} + \mathbf{V} \cdot \nabla s \right\} = 2\mu_1 \mathbf{D} : \nabla \mathbf{V} + \lambda_1 (\nabla \cdot \mathbf{V})^2 - \nabla \cdot \boldsymbol{\theta} + \sigma^{-1} J^2 \tag{1.2.3-4}$$

$$\mathbf{J} - \nabla \times \mathbf{H} = 0$$

$$\frac{\partial \mathbf{B}}{\partial t} + \nabla \times \mathbf{E} = 0$$

When the electrical conductivity σ is zero, we have $\mathbf{J} = 0$ and $\sigma^{-1} J^2 = 0$, so that the dynamical equations on the one hand and the electromagnetic equations on the other are independent of each other; we thus revert to the situation in classical aerodynamics. When the electrical resistivity σ^{-1} is zero, we have $\mathbf{E} + \mathbf{V} \times \mathbf{B} = 0$ and $\sigma^{-1} J^2 = 0$. In these two extreme cases, the second term of the energy equation vanishes together with the three dissipation coefficients λ_1, μ_1, and λ. Naturally, it is necessary to add to Eqs. (1.2.3-4) equations which express Ohm's law and the laws governing electromagnetic and mechanical behavior.

1.2.4. Equations in Curvilinear Coordinates

For practical applications, it is important to know how to write the equations of magnetofluiddynamics in a system of general curvilinear coordinates. For this purpose, we recall expressions for the gradient, divergence, and curl in such a system.

Starting with a cartesian space referred to three rectangular coordinate axes, 0, x, y, z, we consider an orthogonal triad of surfaces

$$f_1(x, y, z) = \alpha_1, \qquad f_2(x, y, z) = \alpha_2, \qquad f_3(x, y, z) = \alpha_3$$

The parameters α_1, α_2, α_3 are called curvilinear orthogonal coordinates. The correspondence between cartesian coordinates x, y, z

and the curvilinear coordinates α_1, α_2, α_3 is supposed one-to-one, so that the preceding formulas can be expressed in the form

$$x = g_1(\alpha_1, \alpha_2, \alpha_3), \qquad y = g_2(\alpha_1, \alpha_2, \alpha_3), \qquad z = g_3(\alpha_1, \alpha_2, \alpha_3)$$

When α_i $(i = 1, 2, 3)$ varies alone, the point M, with coordinates x, y, z describes the coordinate line α_i, along which the curvilinear length is denoted by s_i. We have

$$ds_i = d\alpha_i \left[\left(\frac{\partial g_1}{\partial \alpha_i}\right)^2 + \left(\frac{\partial g_2}{\partial \alpha_i}\right)^2 + \left(\frac{\partial g_3}{\partial \alpha_i}\right)^2\right]^{1/2} = \frac{d\alpha_i}{h_i(\alpha_1, \alpha_2, \alpha_3)}$$

The coordinate lines are assumed to be ordered so that the triad formed by the positive directions of the tangents $\mathbf{i}_1$, $\mathbf{i}_2$, $\mathbf{i}_3$, respectively is right handed. We wish to recall the following results: The gradient of a scalar function U is given by the expression

$$\nabla U = \mathbf{i}_1 h_1 \frac{\partial U}{\partial \alpha_1} + \mathbf{i}_2 h_2 \frac{\partial U}{\partial \alpha_2} + \mathbf{i}_3 h_3 \frac{\partial U}{\partial \alpha_3}$$

The divergence of a vector $\mathbf{V} = \mathbf{i}_1 v_1 + \mathbf{i}_2 v_2 + \mathbf{i}_3 v_3$ is given by

$$\nabla \cdot \mathbf{V} = h_1 h_2 h_3 \left\{\frac{\partial}{\partial \alpha_1}\left(\frac{v_1}{h_2 h_3}\right) + \frac{\partial}{\partial \alpha_2}\left(\frac{v_2}{h_3 h_1}\right) + \frac{\partial}{\partial \alpha_3}\left(\frac{v_3}{h_1 h_2}\right)\right\}$$

and the curl of the vector $\mathbf{V}$ is determined by the expression

$$\nabla \times \mathbf{V} = h_1 h_2 h_3 \left\{\frac{\mathbf{i}_1}{h_1}\left[\frac{\partial}{\partial \alpha_2}\left(\frac{v_3}{h_3}\right) - \frac{\partial}{\partial \alpha_3}\left(\frac{v_2}{h_2}\right)\right]\right\} + \cdots$$

We shall apply these different formulas to three particular systems of coordinates. We shall restrict ourselves to the case where all the dissipation coefficients λ_1, μ_1, λ, and σ^{-1} are zero or where the external forces $\rho\mathbf{F}$ are negligible.

a. Cartesian Coordinates

We choose a system of cartesian rectangular coordinates x, y, z. We denote by u, v, w the components of the vector $\mathbf{V}$ along the coordinate axis and those of the vector $\mathbf{H}$ by α, β, γ. The equations of motion are written in the following form:

$$\frac{\partial u}{\partial t} + u\frac{\partial u}{\partial x} + v\frac{\partial u}{\partial y} + w\frac{\partial u}{\partial z} + \frac{1}{\rho}\frac{\partial p}{\partial x} = \frac{\mu}{\rho} A$$

$$\frac{\partial v}{\partial t} + u\frac{\partial v}{\partial x} + v\frac{\partial v}{\partial y} + w\frac{\partial v}{\partial z} + \frac{1}{\rho}\frac{\partial p}{\partial y} = \frac{\mu}{\rho}B$$

$$\frac{\partial w}{\partial t} + u\frac{\partial w}{\partial x} + v\frac{\partial w}{\partial y} + w\frac{\partial w}{\partial z} + \frac{1}{\rho}\frac{\partial p}{\partial z} = \frac{\mu}{\rho}C$$

$$\frac{\partial \rho}{\partial t} + \frac{\partial(\rho u)}{\partial x} + \frac{\partial(\rho v)}{\partial y} + \frac{\partial(\rho w)}{\partial z} = 0$$

$$\frac{\partial s}{\partial t} + u\frac{\partial s}{\partial x} + v\frac{\partial s}{\partial y} + w\frac{\partial s}{\partial z} = 0$$

$$\frac{\partial \alpha}{\partial t} - \frac{\partial}{\partial y}(u\beta - v\alpha) + \frac{\partial}{\partial z}(w\alpha - u\gamma) = 0$$

$$\frac{\partial \beta}{\partial t} - \frac{\partial}{\partial z}(v\gamma - w\beta) + \frac{\partial}{\partial x}(u\beta - v\alpha) = 0$$

$$\frac{\partial \gamma}{\partial t} - \frac{\partial}{\partial x}(w\alpha - u\gamma) + \frac{\partial}{\partial y}(v\gamma - w\beta) = 0$$

with

$$A = \beta\frac{\partial \alpha}{\partial y} + \gamma\frac{\partial \alpha}{\partial z} - \beta\frac{\partial \beta}{\partial x} - \gamma\frac{\partial \gamma}{\partial x}$$

$$B = \gamma\frac{\partial \beta}{\partial z} + \alpha\frac{\partial \beta}{\partial x} - \gamma\frac{\partial \gamma}{\partial y} - \alpha\frac{\partial \alpha}{\partial y}$$

$$C = \alpha\frac{\partial \gamma}{\partial x} + \beta\frac{\partial \gamma}{\partial y} - \alpha\frac{\partial \alpha}{\partial z} - \beta\frac{\partial \beta}{\partial z}$$

b. *Spherical Coordinates*

We make the change of variables

$$x = r\sin\theta\cos\varphi, \qquad 0 \leqslant \varphi < 2\pi$$

$$y = r\sin\theta\sin\varphi, \qquad 0 \leqslant \theta \leqslant \pi$$

$$z = r\cos\theta$$

In terms of the notation given above, we obtain

$$\alpha_1 = r, \qquad \alpha_2 = \theta, \qquad \alpha_3 = \varphi$$

$$h_1 = 1, \qquad h_2 = 1/r, \qquad h_3 = 1/(r\sin\theta)$$

We denote by u, v, w the components of the vector $\mathbf{V}$ along the axes $\mathbf{i}_1$, $\mathbf{i}_2$, $\mathbf{i}_3$ associate with spherical coordinates and the components of the vector $\mathbf{H}$ along the same axes by α, β, γ. The equations of motion can be written in the following form:

$$\frac{\partial u}{\partial t} + u\frac{\partial u}{\partial r} + \frac{v}{r}\frac{\partial u}{\partial \theta} + \frac{w}{r\sin\theta}\frac{\partial u}{\partial \varphi} + \frac{1}{\rho}\frac{\partial p}{\partial r} - \frac{v^2 + w^2}{r} = \frac{\mu}{\rho}A$$

$$\frac{\partial v}{\partial t} + u\frac{\partial v}{\partial r} + \frac{v}{r}\frac{\partial v}{\partial \theta} + \frac{w}{r\sin\theta}\frac{\partial v}{\partial \varphi} + \frac{1}{\rho}\frac{\partial p}{\partial \theta} + \frac{uv - w^2\cot\theta}{r} = \frac{\mu}{\rho}B$$

$$\frac{\partial w}{\partial t} + u\frac{\partial w}{\partial r} + \frac{v}{r}\frac{\partial w}{\partial \theta} + \frac{w}{r\sin\theta}\frac{\partial w}{\partial \varphi} + \frac{1}{r\sin\theta}\frac{\partial p}{\partial \varphi} + \frac{uw + vw\cot\theta}{r} = \frac{\mu}{\rho}C$$

$$r^2\sin\theta\frac{\partial \rho}{\partial t} + \frac{\partial}{\partial r}(r^2\rho u\sin\theta) + \frac{\partial}{\partial \theta}(r\rho v\sin\theta) + \frac{\partial}{\partial \varphi}(r\rho w) = 0$$

$$\frac{\partial s}{\partial t} + u\frac{\partial s}{\partial r} + \frac{v}{r}\frac{\partial s}{\partial \theta} + \frac{w}{r\sin\theta}\frac{\partial s}{\partial \varphi} = 0$$

$$\frac{\partial \alpha}{\partial t} + \frac{1}{r^2\sin\theta}\left\{\frac{\partial}{\partial \varphi}[r(w\alpha - u\gamma)] - \frac{\partial}{\partial \theta}[r\sin\theta(u\beta - v\alpha)]\right\} = 0$$

$$\frac{\partial \beta}{\partial t} + \frac{1}{r\sin\theta}\left\{\frac{\partial}{\partial r}[r\sin\theta(u\beta - v\alpha)] - \frac{\partial}{\partial \varphi}[v\gamma - w\beta]\right\} = 0$$

$$\frac{\partial \gamma}{\partial t} + \frac{1}{r}\left\{\frac{\partial}{\partial \theta}[v\gamma - w\beta] - \frac{\partial}{\partial r}[r(w\alpha - u\gamma)]\right\} = 0$$

with

$$A = -\beta\frac{\partial \beta}{\partial r} - \gamma\frac{\partial \gamma}{\partial r} + \frac{\beta}{r}\frac{\partial \alpha}{\partial \theta} + \frac{\gamma}{r\sin\theta}\frac{\partial \alpha}{\partial \varphi} - \frac{\beta^2 + \gamma^2}{r}$$

$$B = -\frac{\alpha}{r}\frac{\partial \alpha}{\partial \theta} - \frac{\gamma}{r}\frac{\partial \gamma}{\partial \theta} + \alpha\frac{\partial \beta}{\partial r} + \frac{\gamma}{r\sin\theta}\frac{\partial \beta}{\partial \varphi} + \frac{\alpha\beta - \gamma^2\cot\theta}{r}$$

$$C = -\frac{\alpha}{r\sin\theta}\frac{\partial \alpha}{\partial \varphi} - \frac{\beta}{r\sin\theta}\frac{\partial \beta}{\partial \varphi} + \alpha\frac{\partial \gamma}{\partial r} + \frac{\beta}{r}\frac{\partial \gamma}{\partial \theta} + \frac{\alpha\gamma + \beta\gamma\cot\theta}{r}$$

c. Cylindrical Coordinates

We make the change of variables

$$x = x, \qquad y = r\cos\varphi, \qquad z = r\sin\varphi, \qquad 0 \leqslant \varphi < 2\pi$$

and note that

$$\alpha_1 = x, \qquad \alpha_2 = r, \qquad \alpha_3 = \varphi$$

$$h_1 = 1, \qquad h_2 = 1, \qquad h_3 = 1/r$$

We denote by u, v, w the components of the vector **V** along the axes associated with cylindrical coordinates and the components of the vector **H** along these axes by α, β, γ. The equations of motion can be written in the following form:

$$\frac{\partial u}{\partial t} + u\frac{\partial u}{\partial x} + v\frac{\partial u}{\partial r} + \frac{w}{r}\frac{\partial u}{\partial \varphi} + \frac{1}{\rho}\frac{\partial p}{\partial x} = \frac{\mu}{\rho}A$$

$$\frac{\partial v}{\partial t} + u\frac{\partial v}{\partial x} + v\frac{\partial v}{\partial r} + \frac{w}{r}\frac{\partial v}{\partial \varphi} + \frac{1}{\rho}\frac{\partial p}{\partial r} - \frac{w^2}{r} = \frac{\mu}{\rho}B$$

$$\frac{\partial w}{\partial t} + u\frac{\partial w}{\partial x} + v\frac{\partial w}{\partial r} + \frac{w}{r}\frac{\partial w}{\partial \varphi} + \frac{1}{\rho r}\frac{\partial p}{\partial \varphi} + \frac{vw}{r} = \frac{\mu}{\rho}C$$

$$r\frac{\partial \rho}{\partial t} + \frac{\partial}{\partial x}(r\rho u) + \frac{\partial}{\partial r}(r\rho v) + \frac{\partial}{\partial \varphi}(\rho w) = 0$$

$$\frac{\partial s}{\partial t} + u\frac{\partial s}{\partial x} + v\frac{\partial s}{\partial r} + \frac{w}{r}\frac{\partial s}{\partial \varphi} = 0$$

$$\frac{\partial \alpha}{\partial t} + \frac{1}{r}\left\{\frac{\partial}{\partial \varphi}(w\alpha - u\gamma) - \frac{\partial}{\partial r}[r(u\beta - v\alpha)]\right\} = 0$$

$$\frac{\partial \beta}{\partial t} + \frac{1}{r}\left\{\frac{\partial}{\partial x}[r(u\beta - v\alpha)] - \frac{\partial}{\partial \varphi}(v\gamma - w\beta)\right\} = 0$$

$$\frac{\partial \gamma}{\partial t} + \frac{\partial}{\partial r}(v\gamma - w\beta) - \frac{\partial}{\partial x}(w\alpha - u\gamma) = 0$$

with

$$A = -\beta\frac{\partial \beta}{\partial x} - \gamma\frac{\partial \gamma}{\partial x} + \beta\frac{\partial \alpha}{\partial r} + \frac{\gamma}{r}\frac{\partial \alpha}{\partial \varphi}$$

$$B = -\alpha\frac{\partial \alpha}{\partial r} - \gamma\frac{\partial \gamma}{\partial r} + \beta\frac{\partial \beta}{\partial x} + \frac{\gamma}{r}\frac{\partial \beta}{\partial \varphi} - \frac{\gamma^2}{r}$$

$$C = -\frac{\alpha}{r}\frac{\partial \alpha}{\partial \varphi} - \frac{\beta}{r}\frac{\partial \beta}{\partial \varphi} + \alpha\frac{\partial \gamma}{\partial x} + \beta\frac{\partial \gamma}{\partial r} + \frac{\beta\gamma}{r}$$

1.3. The Shock Equations

1.3.1. Preliminary Formulas

To establish the shock equations, we start with the equations of motion written in integral form. In these equations, derivatives with respect to time of integrals of volume occur. We are going to evaluate these derivatives in the case when the function to be integrated is continuous on a surface Σ. The surface Σ varies with time; we denote by dn the distance normal to Σ between the positions occupied at the very close instants t and $t + dt$; the quantity U defined by $dn = U\,dt$ is called the speed of displacement of the surface Σ.

We always consider the volume (v) to be formed by elements of matter which do not mix and its boundary (S). The surface (Σ) divides the volume (v) in two volumes (v_1) and (v_2), bounded by parts (S_1) and (S_2) of the surface (S). We propose to calculate the derivative

$$\frac{d}{dt}\iiint_{(v)} \varphi\, d\tau$$

The scalar function φ has the value φ_1 in the volume (v_1), region (1), and φ_2 in the volume (v_2), region (2). On the surface (Σ), the function is discontinuous so that $[\varphi] = \varphi_2 - \varphi_1 \neq 0$. The speed of displacement of the surface Σ is counted positive when going from region (2) toward region (1). The volume swept out by the area $d\sigma$ of Σ during the time dt is $U\,d\sigma\,dt$; and the variation of the integral $\iiint \varphi\, d\tau$ resulting from the corresponding displacement of area $d\sigma$ has the value

$$(\varphi_2 - \varphi_1)U\,d\sigma\,dt$$

We deduce the following formula, which is a generalization of the first of the formulas (1.2.1-4):

$$\frac{d}{dt}\iiint_{(v)} \varphi\, d\tau = \iiint_{(v_1)} \frac{\partial\varphi}{\partial t}\, d\tau + \iiint_{(v_2)} \frac{\partial\varphi}{\partial t}\, d\tau + \iint_{(\Sigma)} (\varphi_2 - \varphi_1)U\,d\sigma + \iint_{(S)} \varphi(\mathbf{V}\cdot\mathbf{n})\,ds \qquad (1.3.1\text{-}1)$$

The derivative $(\partial\varphi/\partial t)$ is assumed to be bounded. We suppose that the time is fixed, so that the surface Σ is at rest, and we allow the two portions (S_1) and (S_2) of the surface (S) to tend toward (Σ) in such a way that the limiting positions of the normals to (S_1) and (S_2) coincide with the normal to (Σ). In the limit, we obtain the formula

$$\lim_{(S)\to(\Sigma)} \frac{d}{dt} \iiint_{(v)} \varphi \, d\tau = \iint_{(\Sigma)} [\varphi(U - V_n)] \, d\sigma \tag{1.3.1-2}$$

with

$$[\varphi(U - V_n)] = \varphi_2(U - V_{n_2}) - \varphi_1(U - V_{n_1})$$

The scalar product $V_n = \mathbf{V} \cdot \mathbf{n}$ represents the component of the fluid speed along the normal to the surface Σ; $\mathbf{n}$ is the unit normal to (Σ) and directed from region (2) toward region (1). The quantity $U - V_n$ is called the speed of propagation of the surface Σ with respect to the fluid.

When the scalar function φ is replaced by a vector function $\mathbf{A}$, we obtain the following formulas:

$$\frac{d}{dt} \iiint_{(v)} \mathbf{A} \, d\tau = \iiint_{(v_1)} \frac{\partial \mathbf{A}}{\partial t} \, d\tau + \iiint_{(v_2)} \frac{\partial \mathbf{A}}{\partial t} \, d\tau + \iint_{(\Sigma)} (\mathbf{A}_2 - \mathbf{A}_1) U \, d\sigma + \iint_{(S)} \mathbf{A}(\mathbf{V} \cdot \mathbf{n}) \, ds \tag{1.3.1-3}$$

$$\lim_{(S)\to(\Sigma)} \frac{d}{dt} \iiint_{(v)} \mathbf{A} \, d\tau = \iint_{(\Sigma)} [\mathbf{A}(U - V_n)] \, d\sigma \tag{1.3.1-4}$$

with

$$[\mathbf{A}(U - V_n)] = \mathbf{A}_2(U - V_{n_2}) - \mathbf{A}_1(U - V_{n_1})$$

When the volume (v) is fixed, the preceding formulas are still valid provided that the speed $\mathbf{V}$ is zero.

1.3.2. Electromagnetic Equation of Shocks

Shock phenomena can only appear in fluid mechanics or in magnetofluiddynamics in the absence of dissipation phenomena. We shall therefore suppose that $\lambda_1 = \mu_1 = \lambda = \sigma^{-1} = 0$.

Since the electrical resistivity is zero, Ohm's law leads to the relation

$$\mathbf{E} + \mathbf{V} \times \mathbf{B} = 0 \tag{1.3.2-1}$$

To obtain the relations between electromagnetic discontinuities it is necessary to write Faraday's law, Eq. (1.1.2-2), in the following form:

$$\frac{d}{dt}\iiint_{(v)} \mathbf{B}\, d\tau + \iint_{(S)} (\mathbf{n} \times \mathbf{E})\, ds = 0 \qquad (1.3.2\text{-}2)$$

The volume (v) and the surface (S) are fixed. To establish Eq. (1.3.2-2), we choose an arbitrary fixed unit vector $\mathbf{k}$ and we denote by z an abscissa measured along an axis defined by $\mathbf{k}$. We take for the surface (Σ) and contour (c) sections by a plane $z =$ const of this volume (v) and its boundary (S). Starting from Eqs. (1.1.2-2), one can write, successively, replacing $\mathbf{n}$ by $\mathbf{k}$ and integrating with respect to z,

$$\frac{d}{dt}\iint_{(\Sigma)} \mathbf{B} \cdot \mathbf{k}\, d\sigma + \int_{(c)} \mathbf{E} \cdot d\mathbf{l} = 0$$

$$\frac{d}{dt}\iiint_{(v)} \mathbf{B} \cdot \mathbf{k}\, d\tau + \iint_{(S)} \mathbf{E} \cdot \boldsymbol{\tau}\, dl\, dz = 0$$

$$\mathbf{k} \cdot \left\{\frac{d}{dt}\iiint_{(v)} \mathbf{B}\, d\tau + \iint_{(S)} (\mathbf{n} \times \mathbf{E})\, ds\right\} = 0$$

$$\frac{d}{dt}\iiint_{(v)} \mathbf{B}\, d\tau + \iint_{(S)} (\mathbf{n} \times \mathbf{E})\, ds = 0$$

where $\boldsymbol{\tau}$ denotes the unit vector along the tangent to the contour (c); as this contour is in a plane perpendicular to $\mathbf{k}$, we have $\sin\theta\, \boldsymbol{\tau} = \mathbf{k} \times \mathbf{n}$ and $\sin\theta\, \mathbf{E} \cdot \boldsymbol{\tau} = \mathbf{k} \cdot (\mathbf{n} \times \mathbf{E})$. The last relation follows from the fact that the vector $\mathbf{k}$ is arbitrary. To deduce Faraday's law, written in the form (1.3.2-2) corresponding to the shock equation, it is sufficient to apply formula (1.3.1-4) in the case of a fixed volume ($\mathbf{V} = 0$):

$$\lim_{(S)\to(\Sigma)} \frac{d}{dt}\iiint_{(v)} \mathbf{B}\, d\tau = \iint_{(\Sigma)} [\mathbf{B}U]\, d\sigma$$

$$\lim_{(S)\to(\Sigma)} \iint_{(S)} (\mathbf{n} \times \mathbf{E})\, ds = \iint_{(\Sigma)} -\mathbf{n} \times [\mathbf{E}]\, d\sigma$$

It must be noticed that, when $(S) \to (\Sigma)$, $\mathbf{n}_{(S)} \times \mathbf{E}$ has the limiting value $\mathbf{n}_{(\Sigma)} \times \mathbf{E}_1$ in region (1), where the limit of $\mathbf{n}_{(S)}$ is $\mathbf{n}_{(\Sigma)}$, while in region (2), $\mathbf{n}_{(S)} \times \mathbf{E}$ has as its limit $-\mathbf{n}_{(\Sigma)} \times \mathbf{E}_2$. We finally obtain the following shock equation:

$$U\mathbf{B}_2 - \mathbf{n} \times \mathbf{E}_2 = U\mathbf{B}_1 - \mathbf{n} \times \mathbf{E}_1 \qquad (1.3.2\text{-}3)$$

That is,

$$U\mathbf{B}_2 + \mathbf{n} \times (\mathbf{V}_2 \times \mathbf{B}_2) = U\mathbf{B}_1 + \mathbf{n} \times (\mathbf{V}_1 \times \mathbf{B}_1)$$

or

$$(V_{\mathrm{n}_2} - U)\mathbf{B}_2 - B_{\mathrm{n}_2}\mathbf{V}_2 = (V_{\mathrm{n}_1} - U)\mathbf{B}_1 - B_{\mathrm{n}_1}\mathbf{V}_1 \qquad (1.3.2\text{-}4)$$

We have put $V_{\mathrm{n}} = \mathbf{V} \cdot \mathbf{n}$ and $B_{\mathrm{n}} = \mathbf{B} \cdot \mathbf{n}$. If we next put $\mathbf{v} = \mathbf{n} \times (\mathbf{V} \times \mathbf{n})$ and $\mathbf{b} = \mathbf{n} \times \ \mathbf{B} \times \mathbf{n})$, Eq. (1.3.2-4) has the following projections on the normal (taking account of the fact that there always exists a direction along which $U \neq 0$) and on the tangent plane to the shock wave:

$$B_{\mathrm{n}_2} = B_{\mathrm{n}_1}$$

$$(V_{\mathrm{n}_2} - U)\mathbf{b}_2 - B_{\mathrm{n}}\mathbf{v}_2 = (V_{\mathrm{n}_1} - U)\mathbf{b}_1 - B_{\mathrm{n}}\mathbf{v}_1$$

We deduce in particular that a shock wave cannot give rise to a magnetic induction, for $\mathbf{B}_1 = 0$ implies $\mathbf{B}_2 = 0$.

1.3.3. Mechanical Shock Equations

In the absence of dissipation phenomena, the dynamical equations are written in the following integral form:

$$\frac{d}{dt} \iiint_{(v)} \rho \mathbf{V} \, d\tau = \iiint_{(v)} \mathbf{J} \times \mathbf{B} \, d\tau - \iint_{(S)} \mathbf{n} p \, ds$$

$$\frac{d}{dt} \iiint_{(v)} \rho \, d\tau = 0$$

$$\frac{d}{dt} \iiint_{(v)} \rho \left(\epsilon_{\mathrm{i}} + \frac{V^2}{2}\right) d\tau = \iiint_{(v)} \mathbf{J} \cdot \mathbf{E} \, d\tau - \iint_{(S)} p\mathbf{n} \cdot \mathbf{V} \, d\sigma$$

The external forces, which do not appear in the shock equations, since they are bounded, are assumed to be zero. To be able to apply the formula established in Section 1.3.1, and to derive the shock equations, it is necessary to transform the integrals

$$\iiint_{(v)} \mathbf{J} \times \mathbf{B}\, d\tau, \qquad \iiint_{(v)} (\mathbf{J} \times \mathbf{B}) \cdot \mathbf{V}\, d\tau$$

$(\mathbf{J} \times \mathbf{B}) \cdot \mathbf{V}$ being precisely the specific energy of electromagnetic origin $\epsilon_{\mathrm{m}} = \mathbf{J} \cdot \mathbf{E}$, since $\mathbf{E} = -\mathbf{V} \times \mathbf{B}$.

When the electrical resistivity is negligible. Ohm's law does not allow us to determine the electric current density. This indeterminacy requires the introduction of an additional hypothesis related specifically to the calculation of the preceding integrals and which will permit these integrals to be written in a form leading to the shock equations.

In the case of continuous differentiable functions, we have

$$\nabla \times \mathbf{H} = \mathbf{J}, \qquad \nabla \times (\mathbf{V} \times \mathbf{B}) = \partial \mathbf{B}/\partial t$$

At the same time, the following identities are satisfied:

$$\iiint_{(v)} \{(\nabla \times \mathbf{B}) \times \mathbf{B} + \mathbf{B}\, \nabla \cdot \mathbf{B}\}\, d\tau = \iint_{(S)} \mathsf{B} \cdot \mathbf{n}\, ds$$

$$\iiint_{(v)} \{[(\nabla \times \mathbf{B}) \times \mathbf{B}] \cdot \mathbf{V} + [\nabla \times (\mathbf{V} \times \mathbf{B})] \cdot \mathbf{B} + \nabla \cdot \tfrac{1}{2} B^2 \mathbf{V}\}\, d\tau$$
$$= \iint_{(S)} (\mathsf{B} \cdot \mathbf{n}) \cdot \mathbf{V}\, ds$$

with

$$\mathsf{B} = (\mathbf{B}, \mathbf{B}) - (B^2 \mathsf{U}/2)$$

and

$$\mathsf{B} \cdot \mathbf{n} = \mathbf{B}\, B_{\mathrm{n}} - (\mathbf{n}\, B^2/2), \qquad B_{\mathrm{n}} = \mathbf{B} \cdot \mathbf{n}$$

From the last of Eqs. (1.2.3-2), it follows that the scalar $\nabla \cdot \mathbf{B}$ is independent of time; if we suppose that $\nabla \cdot \mathbf{B} = 0$ initially, then $\nabla \cdot \mathbf{B} = 0$ at all times. Using this result and replacing $\nabla \times (\mathbf{V} \times \mathbf{B})$ by $\partial \mathbf{B}/\partial t$, we can write

$$\iiint_{(v)} \nabla \times \mathbf{B} \times \mathbf{B}\, d\tau = \iint_{(S)} \mathsf{B} \cdot \mathbf{n}\, ds$$

$$\iiint_{(v)} [(\nabla \times \mathbf{B}) \times \mathbf{B}] \cdot \mathbf{V}\, d\tau = -\frac{d}{dt} \iiint_{(v)} \frac{B^2\, d\tau}{2} + \iint_{(S)} (\mathsf{B} \cdot \mathbf{n}) \cdot \mathbf{V}\, ds$$

We now suppose that the medium obeys the electromagnetic law $\mathbf{B} = \mu(\rho, T)\,\mathbf{H}$ and that the permeability $\mu(\rho, T)$ can be considered to be constant. The dynamical equations are then written in the form

$$\frac{d}{dt}\iiint_{(v)} \rho\mathbf{V}\, d\tau + \iint_{(S)} \boldsymbol{\pi}\, ds = 0$$

$$\frac{d}{dt}\iiint_{(v)} \rho\, d\tau = 0 \qquad (1.3.3\text{-}1)$$

$$\frac{d}{dt}\iiint_{(v)} \left\{\rho\left(\epsilon_{\mathrm{i}} + \frac{V^2}{2}\right) + \frac{B^2}{2\mu}\right\} d\tau + \iint_{(S)} \boldsymbol{\pi}\cdot\mathbf{V}\, ds = 0$$

with

$$\boldsymbol{\pi} = \mathbf{n}p + \mathbf{n}(\mu H^2/2) - \mu H_{\mathrm{n}}\,\mathbf{H} \qquad (1.3.3\text{-}2)$$

The equations (1.3.3-1), established under the hypothesis of continuous differential functions, no longer contain derivatives. We then require that these equations still be valid in the case of discontinuous functions: this supplementary hypothesis in fact amounts to a law of magnetofluiddynamic shock waves. We then deduce from Eqs. (1.3.3-1) that the following three quantities are continuous across a shock wave:

$$(V_{\mathrm{n}} - U)\rho \qquad (1.3.3\text{-}3)$$

$$(V_{\mathrm{n}} - U)\rho\mathbf{V} + \boldsymbol{\pi} \qquad (1.3.3\text{-}4)$$

$$(V_{\mathrm{n}} - U)\{\rho(\tfrac{1}{2}V^2 + \epsilon_{\mathrm{i}}) + \tfrac{1}{2}\mu H^2\} + \boldsymbol{\pi}\cdot\mathbf{V} \qquad (1.3.3\text{-}5)$$

Adding the condition of continuity of the vector

$$(V_{\mathrm{n}} - U)\mathbf{H} - H_{\mathrm{n}}\mathbf{V} \qquad (1.3.3\text{-}6)$$

we obtain a complete system of shock equations. These equations determine the values of the quantities $\mathbf{H}$, $\mathbf{V}$, p, and ρ behind the shock as a function of the values in front of the shock and of the speed of displacement of the shock wave U.

CHAPTER 2

PROPAGATION OF SMALL DISTURBANCES

The study of the propagation of infinitesimally small disturbances in a continuous medium leads to the discussion of Cauchy's problem. In the case of MFD, the problem is that of a system of eight partial differential equations of hyperbolic type. Section 2.1 will deal with wave surfaces (characteristic manifolds), speeds of propagation, and the derivation of relations satisfied by discontinuities in various quantities across wave surfaces.

The theory of characteristic manifolds will lead to a study of simple waves in Section 2.2; simple waves are the solutions of the equations of motion for which all the unknown quantities depend on single functions of space and time.

2.1. Propagation of Small Disturbances

2.1.1. Discussion of Cauchy's Problem

The study of the propagation of infinitesimally small disturbances in a continuous medium leads to the discussion of a classical mathematical problem called Cauchy's problem. We shall present the problem of propagation of small disturbances in the case of a compressible fluid in the presence of a magnetic field. When viscosity, thermal conductivity, and electrical resistivity are neglected, the motion is determined by the equations

$$\rho \frac{\partial \mathbf{V}}{\partial t} + (\nabla \times \mathbf{V}) \times \mathbf{V} + \nabla \left(\frac{V^2}{2} \right) = -\nabla p + (\nabla \times \mathbf{H}) \times \mu \mathbf{H}$$

$$\frac{\partial \rho}{\partial t} + \nabla \cdot \rho \mathbf{V} = 0$$

$$\frac{\partial S}{\partial t} + \mathbf{V} \cdot \nabla S = 0 \qquad (2.1.1\text{-}1)$$

$$\frac{\partial \mathbf{H}}{\partial t} - \nabla \times (\mathbf{V} \times \mathbf{H}) = 0$$

The unknowns are **H**, **V**, **S**, and ρ; the pressure p is a known function of ρ and S. We denote by (Σ) a manifold of three dimensions in a space in four dimensions t, x, y, z. Cauchy's problem consists of finding a solution of Eqs. (2.1.1-1) which takes the given values on the manifold (Σ).

The sections of (Σ) by planes $t =$ const form a family of surfaces $\Sigma(t)$ in the three-dimensional Euclidean space x, y, z; one surface corresponds to each value of t. The length of the normal segment separating two surfaces corresponding to infinitesimally close times t and $t + dt$ will be called dn; the ratio $U = dn/dt$ represents then the velocity of displacement of the surface $\Sigma(t)$; this velocity is a function of x, y, z, t. On the manifold (Σ) we denote by d^n a displacement normal to the sections $\Sigma(t)$ (Fig. 2.1.1-a).

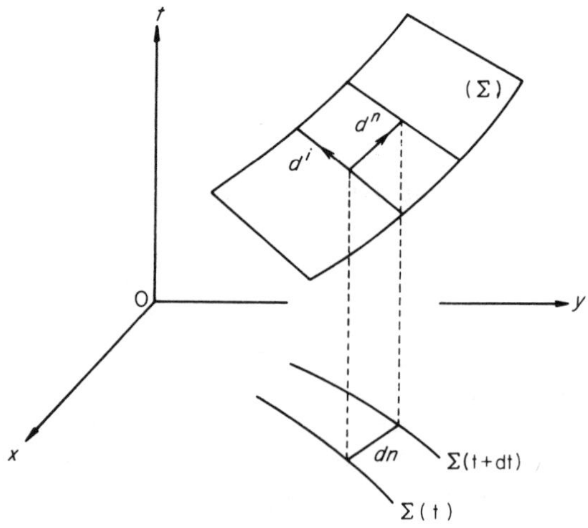

FIG. 2.1.1.a. Displacement d^n normal to the sections $\Sigma(t)$.

The values of the functions **H**, **V**, ρ, and S are given on the manifold (Σ); differentiation of these data on the manifold (Σ) permits us to write 24 scalar relations.

To discuss these scalar relations in detail, it is convenient to introduce a system of orthogonal curvilinear coordinates such that $\Sigma(t)$ form one of the families of coordinate surfaces. These surfaces have the equations

$$f_1(x, y, z) = \alpha_1, \qquad f_2(x, y, z) = \alpha_2, \qquad f_3(x, y, z) = \alpha$$

the last family representing the surfaces $\Sigma(t)$; the parameter α is a function of time, the parameters α_1 and α_2 are independent of time. When displacements are made along the coordinate lines, the curvilinear abscissa has the value s_1, s_2, or s such that

$$ds_1 = \frac{d\alpha_1}{h_1(\alpha_1, \alpha_2, \alpha)}, \qquad ds_2 = \frac{d\alpha_2}{h_2(\alpha_1, \alpha_2, \alpha)}, \qquad ds = dn = \frac{d\alpha}{h(\alpha_1, \alpha_2, \alpha)}$$

where the coefficients h_1, h_2, h_3 are those introduced in Section 1.2.4, replacing the parameter α_3 by α. The scalar relations obtained by differentiating the data on the manifold (Σ) are as follows:

$$\begin{aligned}
d^{\mathrm{n}}H &= h\frac{\partial \mathbf{H}}{\partial \alpha}\,dn + \frac{\partial \mathbf{H}}{\partial t}\,dt, & d^{i}H &= h_i\frac{\partial \mathbf{H}}{\partial \alpha_i}\,ds_i \\
d^{\mathrm{n}}V &= h\frac{\partial \mathbf{V}}{\partial \alpha}\,dn + \frac{\partial \mathbf{V}}{\partial t}\,dt, & d^{i}V &= h_i\frac{\partial \mathbf{V}}{\partial \alpha_i}\,ds_i \\
d^{\mathrm{n}}\rho &= h\frac{\partial \rho}{\partial \alpha}\,dn + \frac{\partial \rho}{\partial t}\,dt, & d^{i}\rho &= h_i\frac{\partial \rho}{\partial \alpha_i}\,ds_i \\
d^{\mathrm{n}}S &= h\frac{\partial S}{\partial \alpha}\,dn + \frac{\partial S}{\partial t}\,dt, & d^{i}S &= h_i\frac{\partial S}{\partial \alpha_i}\,ds_i
\end{aligned} \tag{2.1.1-2}$$

In these relations dn denotes the normal element, already introduced, between two infinitesimally close surfaces $\Sigma(t)$ and $\Sigma(t + dt)$; ds_i represents the curved abscissa for elementary displacements tangent to $\Sigma(t)$; the index i takes the value 1 and 2. Since $dn = U\,dt$, we can write

$$d^{\mathrm{n}} = \left(\frac{\partial}{\partial t} + Uh\frac{\partial}{\partial \alpha}\right) dt$$

Relations (2.1.1-2), when associated with the equations of motion (2.2.1-1) determine the following first-order differential coefficients:

$$\partial/\partial t, \quad \partial/\partial \alpha, \quad \partial/\partial \alpha_1 , \quad \partial/\partial \alpha_2$$

Since there are eight scalar unknowns, there are 32 differential coefficients, which satisfy a system of 32 linear algebraic equations. When the determinant of the coefficients of the unknowns is not zero, the system admits one solution and one only. The calculation of the differential coefficients of higher order then shows that the Cauchy problem possesses one and only one solution. When the determinant of the coefficients of the unknowns is zero, Cauchy's problem is either impossible or indeterminate.

To explain the results, it is necessary to introduce a curvilinear system of orthogonal coordinates α, α_1, α_2. Equations (2.1.1-1) are transformed with the help of the following formulas, in which $\partial/\partial n$ is written instead of $\partial/\partial \alpha$, to indicate that the corresponding derivatives are normal derivatives:

$$\nabla p = \mathbf{n}\, h \frac{\partial p}{\partial n} + \boldsymbol{\tau}_1\, h_1 \frac{\partial p}{\partial \alpha_1} + \boldsymbol{\tau}_2\, h_2 \frac{\partial p}{\partial \alpha_2}$$

$$\nabla \cdot \mathbf{V} = hh_1h_2 \left\{ \frac{\partial}{\partial n} \left(\frac{V_n}{h_1h_2} \right) + \frac{\partial}{\partial \alpha_1} \left(\frac{V_1}{hh_2} \right) + \frac{\partial}{\partial \alpha_2} \left(\frac{V_2}{hh_1} \right) \right\}$$

$$\begin{aligned} \nabla \times \mathbf{V} = {} & h_1h_2\mathbf{n} \left\{ \frac{\partial}{\partial \alpha_1} \left(\frac{V_2}{h_2} \right) - \frac{\partial}{\partial \alpha_2} \left(\frac{V_1}{h_1} \right) \right\} \\ & + hh_2\boldsymbol{\tau}_1 \left\{ \frac{\partial}{\partial \alpha_2} \left(\frac{V_n}{h} \right) - \frac{\partial}{\partial n} \left(\frac{V_2}{h_2} \right) \right\} \\ & + hh_1\boldsymbol{\tau}_2 \left\{ \frac{\partial}{\partial n} \left(\frac{V_1}{h_1} \right) - \frac{\partial}{\partial \alpha_1} \left(\frac{V_n}{h} \right) \right\} \end{aligned}$$

The vectors $\mathbf{n}$, $\boldsymbol{\tau}_1$, $\boldsymbol{\tau}_2$ are unit vectors along the directions α, α_1, α_2, respectively.

With the differential Eqs. (2.1.1-1) written in the curvilinear system of coordinates α, α_1, α_2, the differential coefficients $\partial/\partial \tau_i$ can be eliminated using Eqs. (2.1.1-2). Calling $\mathbf{z}$ the vector with components H_n, H_1, H_2, V_n, V_1, V_2, ρ, S, we get a relation between the differential coefficients $\partial \mathbf{z}/\partial t$ and $\partial \mathbf{z}/\partial n$

$$\frac{\partial \mathbf{z}}{\partial t} + hA \frac{\partial \mathbf{z}}{\partial n} = \mathbf{w}_1 \qquad (2.1.1\text{-}3)$$

where A is the matrix

$$A = \begin{bmatrix} 0 & 0 & 0 & 0 & 0 & 0 & 0 & 0 \\ -V_1 & V_{\mathrm{n}} & 0 & H_1 & -H_{\mathrm{n}} & 0 & 0 & 0 \\ -V_2 & 0 & V_{\mathrm{n}} & H_2 & 0 & -H_{\mathrm{n}} & 0 & 0 \\ 0 & \mu H_1/\rho & \mu H_2/\rho & V_{\mathrm{n}} & 0 & 0 & \rho^{-1}\,\partial p/\partial \rho & \rho^{-1}\,\partial p/\partial s \\ 0 & -\mu H_{\mathrm{n}}/\rho & 0 & 0 & V_{\mathrm{n}} & 0 & 0 & 0 \\ 0 & 0 & -\mu H_{\mathrm{n}}/\rho & 0 & 0 & V_{\mathrm{n}} & 0 & 0 \\ 0 & 0 & 0 & -\rho & 0 & 0 & V_{\mathrm{n}} & 0 \\ 0 & 0 & 0 & 0 & 0 & 0 & 0 & V_{\mathrm{n}} \end{bmatrix}$$

while the vector $\mathbf{w}_1$ is a known vector can be expressed as a function of the differentials d^i of the data on the manifold (Σ). Now, if we eliminate the derivatives $\partial \mathbf{z}/\partial t$ using the relation

$$d^{\mathrm{n}}\mathbf{z} = \left(\frac{\partial \mathbf{z}}{\partial t} + Uh\,\frac{\partial \mathbf{z}}{\partial n}\right) dt$$

we get for the vector $\partial \mathbf{z}/\partial n$ the linear equation

$$(A - UI)\frac{\partial \mathbf{z}}{\partial n} = \mathbf{w}_2 \tag{2.1.1-4}$$

where I is the unit matrix and $\mathbf{w}_2$ a known vector.

Cauchy's problem will have one and only one solution if the determinant

$$\Delta = \det(A - UI)$$

is different from zero; Δ is a determinant of the eighth order. When the determinant Δ is equal to zero, Cauchy's problem is either impossible or indeterminate; if it is indeterminate, the data should satisfy certain additional conditions. In this case, the manifold (Σ) is called a characteristic manifold, and the values of $\mathbf{H}$, $\mathbf{V}$, ρ, and S on (Σ) are called characteristic data. The surfaces $\Sigma(t)$ associated with a characteristic manifold (Σ), i.e., the sections of the manifold (Σ) by the planes $t =$ const, are called wave surfaces or, more briefly, waves.

If two different solutions of Eqs. (2.1.1-1) take the same value on a manifold (Σ), this must be a characteristic manifold. On the manifold (Σ), the two solutions can have different first derivatives; in this case, the discontinuities of the derivatives $\partial/\partial n$ across a wave surface $\Sigma(t)$ will not be independent, but will obey certain relations which will be established in Section 2.1.3. When an infinitesimally small discontinuity is propagated in a fluid in motion, we can consider that it influences only the derivatives $\partial/\partial t$, $\partial/\partial x$, $\partial/\partial y$, and $\partial/\partial z$. The locus of the points influenced by the discontinuity is a three-dimensional manifold of the space t, x, y, z on which two solutions of Eqs. (2.1.1-1) take the same value and have different first-order partial derivatives; such a manifold is hence a characteristic manifold. The speed of propagation of the discontinuity is then U; for an observer moving with the fluid, the speed of propagation of the discontinuity is equal to $U - V_n$; this is called the speed of propagation of the discontinuity.

2.1.2. Calculation of Propagation Speeds

Since we can always find a frame of reference in which U is different from zero, the condition $\Delta = 0$ is an algebraic equation of the seventh degree for the unknown $U - V_n$, which represents the speed of propagation (the speed in a frame moving with the fluid) of the surfaces $\Sigma(t)$. The equation $\Delta = 0$ splits into the three equations

$$U - V_n = 0 \quad (2.1.2\text{-}1)$$

$$\rho(U - V_n)^2 - \mu H_n^2 = 0 \quad (2.1.2\text{-}2)$$

$$\frac{\partial p}{\partial \rho}\{\rho(U - V_n)^2 - \mu H_n^2\} - (U - V_n)^2\{\rho(U - V_n)^2 - \mu H^2\} = 0 \quad (2.1.2\text{-}3)$$

The velocities of propagation can always be taken positive or zero, because, for two equal and opposite values, there are two surfaces $\Sigma(t)$ propagating (locally) in opposite directions but at the same speed. Hence, at a certain time and position, four speeds of propagation correspond to each direction. One direction is specified by the unit vector $\mathbf{n}$, and we call θ the angle that $\mathbf{n}$ makes with the direction of the magnetic field. One of the velocities of propagation is zero; the others have the (positive) values U_A, U_-, and U_+ defined by the formulas:

$$U_A^2 = \frac{\mu H^2}{\rho}\cos^2\theta$$

$$2U_-^2 = \frac{\partial p}{\partial \rho} + \frac{\mu H^2}{\rho} - \left\{\left(\frac{\partial p}{\partial \rho} + \frac{\mu H^2}{\rho}\right)^2 - 4\frac{\partial p}{\partial \rho}\frac{\mu H^2}{\rho}\cos^2\theta\right\}^{1/2} \quad (2.1.2\text{-}4)$$

$$2U_+^2 = \frac{\partial p}{\partial \rho} + \frac{\mu H^2}{\rho} + \left\{\left(\frac{\partial p}{\partial \rho} + \frac{\mu H^2}{\rho}\right)^2 - 4\frac{\partial p}{\partial \rho}\frac{\mu H^2}{\rho}\cos^2\theta\right\}^{1/2}$$

The speeds of propagation are called: U_A, the magnetohydrodynamic speed, sometimes called Alfvén speed; U_-, the slow magnetoacoustic speed; and U_+, the fast magnetoacoustic speed. The speeds of propagation satisfy the inequalities

$$U_- \leqslant (\partial p/\partial \rho)^{1/2} \leqslant U_+$$

$$U_- \leqslant U_A \leqslant U_+$$

The speed of propagation at a point is not isotropic, but is axisymmetric with respect to the direction of magnetic field. To represent these velocities, we can construct on the straight line *On*, defining the direcuon of propagation, three points P_A, P_-, and P_+ such that

$$OP_A = U_A, \qquad OP_- = U_-, \qquad OP_+ = U_+$$

The locus of the points P_A consists of two spheres; the locus of P_- and P_+ is an algebraic surface of the sixth degree, a surface of revolution the meridian section of which is shown for different cases in Fig. 2.1.2a. When the magnetic field is zero, we have

$$U_- = U_A = 0, \qquad U_+ = (\partial p/\partial \rho)^{1/2}$$

When the magnetic field is different from zero, the speeds of propagation take special values depending on whether the direction of propagation is perpendicular or parallel to the direction of the magnetic field.

(1) For propagation perpendicular to the magnetic field ($\theta \equiv \pi/2$)

$$U_- = U_A = 0$$

$$U_+^2 = (\partial p/\partial \rho) + (\mu H^2/\rho)$$

(2) For propagation parallel to the magnetic field ($\theta = 0$)

$$U_A^2 = \mu H^2/\rho$$

$$U_-^2 = \inf\{\mu H^2/\rho, \partial p/\partial \rho\}$$

$$U_+^2 = \sup\{\mu H^2/\rho, \partial p/\partial \rho\}$$

We can associate with the values of the speed of propagation of

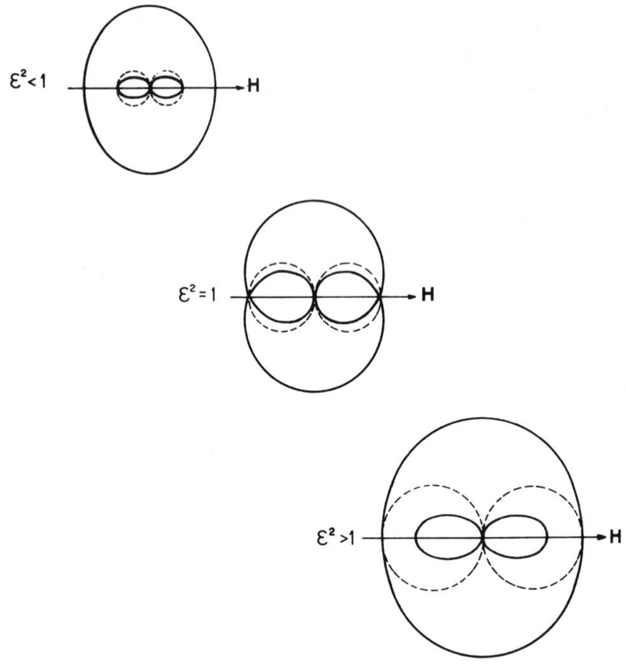

FIG. 2.1.2.a. Speed of propagation of linearized waves. Here, $\epsilon^2 = \mu H^2 \rho / (\partial p / \partial \rho)$. Dashed curve: Alfvén wave; solid curves: magnetoacoustic waves.

small disturbances the corresponding values of the speed of displacement of these perturbations; we get the following values:

$$U = V_{\mathrm{n}}, \qquad U = V_{\mathrm{n}} + U_{\mathrm{A}}, \qquad U = V_{\mathrm{n}} - U_{\mathrm{A}}$$

$$U = V_{\mathrm{n}} + U_{-}, \quad U = V_{\mathrm{n}} - U_{-}, \quad U = V_{\mathrm{n}} + U_{+}, \quad U = V_{\mathrm{n}} - U_{+}$$

These are the eigenvalues, not identically zero, of the matrix A. If the frame of reference moves with the fluid, the velocity V_{n} is zero and the eigenvalue $U = 0$ is then an eigenvalue of order two.

A particularly important case of MFD waves is that of plane waves. We consider plane waves which move parallel to themselves, and we call $\mathbf{n}$ the unit vector perpendicular to these waves, i.e., parallel to the direction of propagation. Each wave passing through a point O at some time taken as the initial instant will, at a later time τ, be at a distance from O (assumed to be moving with the fluid) equal to one of the three values τU_{A}, τU_{-}, τU_{+}; these values depend on the direction **n**. The waves located at distance τU_A pass through one of the two points A and A' such that

$$\mathbf{OA} = (\mu/\rho)^{1/2}\,\mathbf{H}, \qquad \mathbf{OA'} = -(\mu/\rho)^{1/2}\,\mathbf{H}$$

The waves at distance τU_- or τU_+ are tangent to some surface of revolution with the magnetic field direction as axis. The section of this surface is represented on Fig. 2.1.2.b (for different values of the

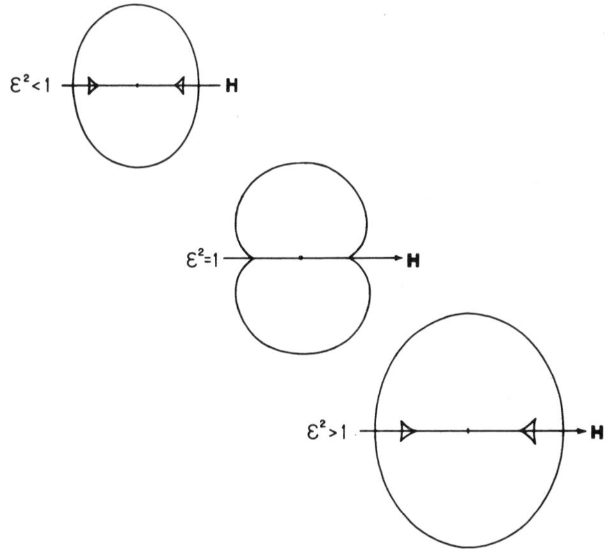

FIG. 2.1.2.b. Friedrichs diagrams. Here, $\epsilon^2 = \mu H^2 \rho/(\partial p/\partial \rho)$.

magnetic field). These curves, called Friedrichs diagrams, are related to the curves of Fig. 2.1.2.a by a simple geometric transformation.

2.1.3. Calculation of the Discontinuities

When the manifold (Σ) is a characteristic manifold for the system of Eqs. (2.1.1-1), Cauchy's problem is either impossible or indeterminate. In the latter case, the general solution of Equation (2.1.1-4) is obtained by adding to a particular solution $(\partial \mathbf{z}/\partial n)^*$ the general solution $\lambda \mathbf{r}$ of the equation with the second term missing,

$$\frac{\partial \mathbf{z}}{\partial n} = \left(\frac{\partial \mathbf{z}}{\partial n}\right)^* + \lambda\,\mathbf{r} \qquad (2.1.3\text{-}1)$$

We have

$$(A - UI)\mathbf{r} = 0 \tag{2.1.3-2}$$

or

$$A\mathbf{r} = U\mathbf{r}$$

so that $\mathbf{r}$ is an eigenvector associated with the eigenvalue U.

When a surface $\Sigma(t)$ is propagated in the fluid, the parameters defining the state of the fluid are continuous. It follows that on a surface $\Sigma(t)$ the tangential derivatives of these quantities are also continuous, while the second derivatives may be discontinuous. Upstream of the surface $\Sigma(t)$, region (1), and downstream of it, region (2), we have

$$(\partial \mathbf{z}/\partial n)_1 = (\partial \mathbf{z}/\partial n)^* + \lambda_1 \mathbf{r}$$

$$(\partial \mathbf{z}/\partial n)_2 = (\partial \mathbf{z}/\partial n)^* + \lambda_2 \mathbf{r}$$

respectively, and hence

$$(\partial \mathbf{z}/\partial n)_2 - (\partial \mathbf{z}/\partial n)_1 = (\lambda_2 - \lambda_1)\mathbf{r} \tag{2.1.3-3}$$

The discontinuity of the normal derivative of the vector $\mathbf{z}$ is proportional to the eigenvector $\mathbf{r}$ of the matrix A, an eigenvector associated with the eigenvalue U, equal to the velocity of displacement of $\Sigma(t)$. To determine these eigenvectors, it is sufficient to consider three cases corresponding to the speeds of propagation U_- and U_+ (treated simultaneously), U_A, and 0.

a. Magnetoacoustic Waves

We consider the case of fast magnetoacoustic waves, which correspond to eigenvalues $V_n + U_+$ and $V_n - U_+$ of the matrix A. We consider the first case: $U = V_n + U_+$; for the second case, we have only to change U_+ to $-U_+$ to find the corresponding eigenvector $V_n - U_+$. The case of slow magnetoacoustic waves can be deduced from that for fast magnetoacoustic waves by interchanging U_+ and U_-.

For $U = V_n + U_+$, the components of the eigenvector $\mathbf{r}$ are

$$\mathbf{r} = \left(0, \frac{\mu H_1 U_+^2}{\rho U_+^2 - \mu H_n^2}, \frac{\mu H_2 U_+^2}{\rho U_+^2 - \mu H_n^2}, \frac{U_+}{\rho},\right.$$

$$\left.\frac{-\mu H_1 H_n U_+}{\rho^2 U_+^2 - \mu\rho H_n^2}, \frac{-\mu H_2 H_n U_+}{\rho^2 U_+^2 - \mu\rho H_n^2}, 1, 0\right)$$

This means that the normal derivatives of **H**, **V**, ρ, and S across a fast magnetoacoustic wave are proportional to the following quantities:

$$\frac{\partial \mathbf{H}}{\partial n} = \frac{\mu(\mathbf{H} - \mathbf{n}H_n)}{\rho U_+^2 - \mu H_n^2} U_+^2$$

$$\frac{\partial \mathbf{V}}{\partial n} = \frac{\rho U_+^2 \mathbf{n} - \mu H_n \mathbf{H}}{\rho U_+^2 - \mu H_n^2} \frac{U_+}{\rho}$$

$$\frac{\partial \rho}{\partial n} = 1, \qquad \frac{\partial S}{\partial n} = 0$$

We deduce that the normal derivative of the pressure,

$$\frac{\partial p}{\partial n} = \left(\frac{\partial p}{\partial \rho}\right)\frac{\partial \rho}{\partial n} + \left(\frac{\partial p}{\partial S}\right)\frac{\partial S}{\partial n}$$

is actually discontinuous. For this reason, waves propagating with one of the velocities U_- or U_+ are called magnetoacoustic waves.

b. Magnetohydrodynamic Waves—Alfvén Waves

The magnetohydrodynamic waves correspond to the eigenvalues $V_n + U_A$ and $V_n - U_A$ of the matrix A. We will treat the first case (the second follows when U_A is replaced by $-U_A$). For $U = V_n + U_A$, the components of the eigenvector $\mathbf{r}$ are

$$\mathbf{r} = (0, (\mu\rho)^{1/2} H_2, -(\mu\rho)^{1/2} H_1, 0, -\mu H_2, \mu H_1, 0, 0)$$

This means that the normal derivatives of the quantities **H**, **V**, ρ, and S across a magnetohydrodynamic wave, of the type we are investigating, are proportional to the quantities

$$\partial\mathbf{H}/\partial n = -(\mu\rho)^{1/2}\, \mathbf{n} \times \mathbf{H}$$

$$\partial\mathbf{V}/\partial n = \mu \mathbf{n} \times \mathbf{H}$$

$$\partial\rho/\partial n = 0, \qquad \partial S/\partial n = 0$$

Since the normal derivative of the density is continuous across the waves, which propagate with speed U_A, these waves are called magnetohydrodynamic waves. The magnetohydrodynamic waves were discovered by Alfvén in 1940 in the case of incompressible fluids; we have established that they exist also for compressible fluids; they are often called Alfvén waves. Across an Alfvén wave, the only variable normal derivatives are those of the tangential magnetic field **h** and tangential velocity **v**. The normal derivative of the square of the tangential magnetic field h^2 is continuous. We have

$$H_n \, \partial\mathbf{v}/\partial n + U_A \, \partial\mathbf{h}/\partial n = 0, \qquad \partial h^2/\partial n = 0$$

c. *Entropy Waves*

Entropy waves correspond to the eigenvalue $U = V_n$ of the matrix A. If we assume $V_n \neq 0$, $\rho \neq 0$, and $H_n \neq 0$, which correspond to the general case, the eigenvector $\mathbf{r}$ corresponding to the eigenvalue $U = V_n$ has the components

$$\mathbf{r} = (0, 0, 0, 0, 0, 0, -(\partial p/\partial S)/(\partial p/\partial \rho), 1)$$

The normal derivatives of the vectors **H** and **V** are zero, the normal derivatives of the quantities ρ and S are proportional to the quantities

$$\frac{\partial \rho}{\partial n} = -\frac{\partial p/\partial S}{\partial p/\partial \rho}, \qquad \frac{\partial S}{\partial n} = 1$$

The corresponding waves are the only ones for which the normal derivative of the specific entropy is discontinuous; for this reason, they are called entropy waves. Entropy waves also exist in aerodynamics, but in that case, the normal derivative of the tangential velocity may be discontinuous.

2.1.4. Equations of Weak Shocks

Extremely weak shock waves are very similar to acoustic waves and the weak shock equations reduce to those which determine normal discontinuities across a magnetoacoustic wave. To verify this result, we recall that the following quantities are invariant across a shock:

$$(V_n - U)\,\mathbf{H} - H_n\mathbf{V}, \qquad (V_n - U)\,\rho\mathbf{V} + \boldsymbol{\pi}, \qquad (V_n - U)\rho$$
$$(V_n - U)\,\rho\{\epsilon_i + \tfrac{1}{2}V^2 + (\mu H^2/2\rho)\} + \boldsymbol{\pi}\cdot\mathbf{V} \tag{2.1.4-1}$$

with

$$\boldsymbol{\pi} = \mathbf{n}(p + \tfrac{1}{2}\mu H^2) - \mu H_n\mathbf{H}$$

When the quantities **H**, **V**, ρ, and S have infinitesimally small discontinuities $\delta\mathbf{H}$, $\delta\mathbf{V}$, $\delta\rho$, and δS across a shock, we can write the first three equations of (2.1.4-1) in the following form (obtained by neglecting products of infinitely small quantities):

$$\begin{aligned} -U\,\delta H_{\mathrm{n}} &= 0 \\ (V_{\mathrm{n}} - U)\,\delta H_i + H_i\,\delta V_{\mathrm{n}} - H_{\mathrm{n}}\,\delta V_i - V_i\,\delta H_{\mathrm{n}} &= 0 \\ (V_{\mathrm{n}} - U)\rho\,\delta V_{\mathrm{n}} + \delta p + \mu(H_1\,\delta H_1 + H_2\,\delta H_2) &= 0 \\ (V_{\mathrm{n}} - U)\rho\,\delta V_i - H_{\mathrm{n}}\,\delta H_i &= 0 \\ (V_{\mathrm{n}} - U)\,\delta\rho + \rho\,\delta V_{\mathrm{n}} &= 0 \end{aligned} \tag{2.1.4-2}$$

where the index i takes the values 1 and 2.

To write the last equation of (2.1.4-1) in the new form, we introduce the specific entropy and the absolute temperature; we have (formula 1.2.2-3)

$$T\,dS = d\epsilon_{\mathrm{i}} + p\,d(1/\rho)$$

The last equation of (2.1.4-1) becomes

$$\begin{aligned} &(V_{\mathrm{n}} - U)\rho\{\mathbf{V}\cdot\delta\mathbf{V} + T\,dS + (p\,\delta\rho/\rho^2)\} + (V_{\mathrm{n}} - U)\mu\mathbf{H}\cdot\delta\mathbf{H} + \delta V_{\mathrm{n}}(p + \mu H^2) \\ &\quad - V_{\mathrm{n}}\rho(V_{\mathrm{n}} - U)\,\delta V_{\mathrm{n}} - \mu H_{\mathrm{n}}(\mathbf{H}\cdot\delta\mathbf{V} + \mathbf{V}\cdot\delta\mathbf{H}) = 0 \end{aligned}$$

The terms containing p as a factor vanish from the last equation of (2.1.4-2). The terms containing $\mathbf{V}$ as a factor are

$$\mathbf{V}\cdot\{\rho(V_{\mathrm{n}} - U)\,\delta\mathbf{V} - \mu H_{\mathrm{n}}\,\delta\mathbf{H}\} = \mathbf{V}\cdot\mathbf{n}\rho(V_{\mathrm{n}} - U)\,\delta V_{\mathrm{n}}$$

The terms containing $\mathbf{H}$ as a factor are

$$\mu\mathbf{H}\cdot\{(V_{\mathrm{n}} - U)\,\delta\mathbf{H} - H_{\mathrm{n}}\,\delta\mathbf{V}\} = \mu\mathbf{H}\cdot(-\mathbf{H}\,\delta V_{\mathrm{n}})$$

All of these terms disappear, together with those containing δV_{n} as a factor. We finally get

$$\rho(V_{\mathrm{n}} - U)T\,\delta S = 0 \tag{2.1.4-3}$$

Equations (2.1.4-2) and (2.1.4-3) combine to yield

$$(A - IU)\,\delta\mathbf{z} = 0 \tag{2.1.4-4}$$

We recover Eq. (2.1.1-4) with the right-hand side zero, and the vector $(\partial\mathbf{z}/\partial n)$ replaced by the discontinuity $\delta\mathbf{z}$. Equation (2.1.4-4) has a nonzero solution only in the case when the determinant

$$\Delta = \det(A - UI)$$

is equal to zero. This means that the extremely weak shock waves propagate with one of the four velocities 0, U_A, U_-, or U_+. Equation (2.1.4-4) can be solved as in the previous section. In the case of magnetofluiddynamic or Alfvén waves, the tangential magnetic field **h** has the same magnitude but can have an arbitrary direction behind the shock; then the shock could not be extremely weak. The same applies to entropy waves when the normal velocity V_n is zero. Therefore, extremely weak shocks are, in general, magnetoacoustic waves.

2.2. Flows with Simple Waves

2.2.1. General Theory of Simple Waves

The equations of MFD have solutions in which all the unknowns are expressed in terms of the variables by means of a single function; the corresponding flows are called simple-wave flows. We will study these motions in the case of one space variable—one-dimensional flows; to begin with, we will develop a general theory of simple waves.

We consider a partial differential equation

$$\partial \mathbf{z}/\partial t + A(\mathbf{z})\, \partial \mathbf{z}/\partial x = 0 \tag{2.2.1-1}$$

where the unknown is a vector $\mathbf{z}$ of Euclidean space in n dimensions, and the matrix A is a given function of $\mathbf{z}$. The system is assumed to be hyperbolic, i.e., the matrix A has n real eigenvalues which we call $g_i(\mathbf{z})$ ($i = 1, 2, \ldots, n$). We call $\mathbf{r}_i(\mathbf{z})$ the eigenvector corresponding to $g_i(\mathbf{z})$; this eigenvector will be unique if the eigenvalue is simple. We have

$$A\mathbf{r}_i(\mathbf{z}) = g_i \mathbf{r}_i(\mathbf{z})$$

The simple-wave solutions of Eq. (2.2.1-1) are solutions of the form $\mathbf{z}(t, x) = \mathbf{z}(X)$ where $X = X(t, x)$. We will have

$$\frac{d\mathbf{z}}{dX}\frac{\partial X}{\partial t} + A(\mathbf{z})\frac{d\mathbf{z}}{dX}\frac{\partial X}{\partial x} = 0$$

or

$$A\frac{d\mathbf{z}}{dX} = -\frac{\partial X/\partial t}{\partial X/\partial x}\frac{d\mathbf{z}}{dX}$$

which proves that the derivative $d\mathbf{z}/dX$ is an eigenvector of the matrix $A(\mathbf{z})$ associated with the eigenvalue

$$g_i(\mathbf{z}) = -\frac{\partial X/\partial t}{\partial X/\partial x}$$

The function $\mathbf{z}(X)$ is determined by integration of the equation

$$d\mathbf{z}/dX = \mathbf{r}_i(\mathbf{z})$$

i.e., of the system

$$\frac{dz_1}{r_i^1(\mathbf{z})} = \frac{dz_2}{r_i^2(\mathbf{z})} = \frac{dz_3}{r_i^3(\mathbf{z})} = \cdots = \frac{dz_n}{r_i^n(\mathbf{z})} = dX \qquad (2.2.1\text{-}2)$$

Once the function $\mathbf{z}(X)$ is known, we get the function $X(t, x)$ by integration of the equations

$$\partial X/\partial t + g_i\{\mathbf{z}(X)\}\, \partial X/\partial x = 0 \qquad (2.2.1\text{-}3)$$

and so find

$$x = g_i\{\mathbf{z}(X)\}t + f(X) \qquad (2.2.1\text{-}4)$$

where the function $f(X)$ is arbitrary. The curves along which the function $X(t, x)$ is constant are called simple waves; simple waves are the straight lines defined by Eq. (2.2.1-4). On a simple wave, the vector $\mathbf{z}(t, x)$ stays constant, if this vector represents a simple-wave solution of Eq. (2.2.1-1).

2.2.2. The Riemann Invariants

To each eigenvector $\mathbf{r}_i(\mathbf{z})$ of the matrix $A(\mathbf{z})$ we can associate the partial differential equation

$$\mathbf{r}_i(\mathbf{z}) \cdot \nabla\varphi(\mathbf{z}) = 0 \qquad (2.2.2\text{-}1)$$

Any solution $\varphi(\mathbf{z})$ of Eq. (2.2.2-1) is called a Riemann invariant of rank i associated with Eq. (2.2.1-1). The problem of finding the Riemann invariants, i.e., the solutions of Eq. (2.2.2-1), reduces to that of finding the first integrals of the system

$$\frac{dz_1}{r_i^1(\mathbf{z})} = \frac{dz_2}{r_i^2(\mathbf{z})} = \cdots = \frac{dz_n}{r_i^n(\mathbf{z})} \qquad (2.2.2\text{-}2)$$

In fact, we call any function $F(z_1, z_2, \ldots, z_n)$ a first integral if it takes a constant value when we replace $z_1, \ldots, z_n$, expressed as a function of z_1, by a system of solutions of Eq. (2.2.2-2). If a function $F(z_1, z_2, \ldots, z_n)$ remains constant for all solutions of Eq. (2.2.2-1), we have

$$dF = \frac{\partial F_1}{\partial z_1} dz_1 + \cdots + \frac{\partial F}{\partial z_n} dz_n = 0$$

or

$$\frac{\partial F}{\partial z_1} r_i^1(\mathbf{z}) + \cdots + \frac{\partial F}{\partial z_n} r_i^n(\mathbf{z}) = 0$$

which means that $F(z_1, z_2, \ldots, z_n)$ is a solution of Eq. (2.2.2-1). Conversely, if Eq. (2.2.2-1) is satisfied, dF is zero for all solutions of the system (2.2.2-2), hence $F(z_1, z_2, \ldots, z_n)$ is a first integral of this system. The ratios appearing in Eq. (2.2.2-2) have the common value

$$\frac{\lambda_1\, dz_1 + \lambda_2\, dz_2 + \cdots + \lambda_n\, dz_n}{\lambda_1 r_i^1 + \lambda_2 r_i^2 + \cdots + \lambda_n r_i^n}$$

where the factors $\lambda_1, \lambda_2, \ldots, \lambda_n$ are arbitrary functions of $\mathbf{z}$. If it is possible to determine the factors λ_j so that the numerator is a total differential of a function F, while the denominator is identically zero, the function F will be a first integral. Naturally, there are no precise rules for finding such factors.

The connection between the Riemann invariants and the simple-wave solutions of Eq. (2.2.1-1) follows from the following theorem:

Theorem. If the solution of Eq. (2.2.1-1) is a constant in a certain region of the t, x plane, the solution in each adjacent region (separated by a straight-line characteristic) is either a constant or a simple wave.

To prove this theorem, we consider a nonuniform region (R) of the t, x plane adjacent to a uniform region. The boundary separating the two regions is a characteristic C which is a straight line, the slope of which is an eigenvalue, g_k say, of the matrix A. We will prove that in the region (R), Eq. (2.2.2-1), in which $\mathbf{r}_i(\mathbf{z})$ is replaced by $\mathbf{r}_k(\mathbf{z})$, has $n - 1$ solutions $\varphi_s(\mathbf{z})$ which are constants. This will permit us to express the components of the vector $\mathbf{z}$ using one of them, z_1, for example, and to write $\mathbf{z} = \mathbf{z}\{z_1(t, x)\}$, which indeed represents a simple-wave flow.

For this purpose, we denote by $\mathbf{l}_j$ the eigenvector to the left of the matrix $A(\mathbf{z})$ corresponding to the eigenvalue g_j. Taking the scalar product of $\mathbf{l}_j$ with Eq. (2.2.1-1), we get

$$\mathbf{l}_j \cdot \left(\frac{\partial \mathbf{z}}{\partial t} + g_j \frac{\partial \mathbf{z}}{\partial x}\right) = 0 \tag{2.2.2-3}$$

because

$$\mathbf{l}_j \cdot \left(A \frac{\partial \mathbf{z}}{\partial x}\right) = (\mathbf{l}_j A) \cdot \frac{\partial \mathbf{z}}{\partial x} = g_j\, \mathbf{l}_j \cdot \frac{\partial \mathbf{z}}{\partial x}$$

The subscript j takes the values 1, 2,..., n.

The eigenvectors on the right, $\mathbf{r}_k$, and those on the left, $\mathbf{l}_j$, of a matrix A form a biorthogonal sequence, i.e., they satisfy the equations

$$\mathbf{l}_j \cdot \mathbf{r}_k \neq 0 \qquad \text{for} \qquad j \neq k \tag{2.2.2-4}$$

In fact, we have

$$A\mathbf{r}_k = g_k\, \mathbf{r}_k$$
$$\mathbf{l}_j A = g_j\, \mathbf{l}_j$$

or

$$\mathbf{l}_j \cdot (A\mathbf{r}_k) = g_k\, \mathbf{l}_j \cdot \mathbf{r}_k$$
$$(\mathbf{l}_j A) \cdot \mathbf{r}_k = g_j\, \mathbf{l}_j \cdot \mathbf{r}_k$$

The left-hand sides are identical; the right-hand sides are therefore also identical, and, putting $g_j \neq g_k$, as we assume, we get Eq. (2.2.2-4).

Equation (2.2.2-4) should be compared with (2.2.2-1) which defines the function $\varphi_s(\mathbf{z})$ and which can be written

$$\mathbf{r}_k(\mathbf{z}) \cdot \nabla\varphi_s(\mathbf{z}) = 0, \qquad s = 1, 2,..., n - 1$$

We will prove that the solutions $\varphi_s(\mathbf{z})$ of this equation all have constant values in (R). Since the eigenvectors $\mathbf{l}_j$ ($j \neq k$) and the gradients of $\varphi_s(\mathbf{z})$ are orthogonal to the same vector (the eigenvector $\mathbf{r}_k$), we have

$$\mathbf{l}_j = \sum_{s=1}^{n-1} b_{js}\, \nabla\varphi_s\,, \qquad j \neq k,$$

where the coefficients b_{js} are functions of $\mathbf{z}$. Hence, Eq. (2.2.2-3) can be written

$$\sum_{s=1}^{n-1} b_{js} \nabla\varphi_s \cdot \left(\frac{\partial \mathbf{z}}{\partial t} + g_j \frac{\partial \mathbf{z}}{\partial x}\right) = 0, \qquad j \neq k$$

or

$$\sum_{s=1}^{n-1} b_{js} \left(\frac{\partial \varphi_s}{\partial t} + g_j \frac{\partial \varphi_s}{\partial x}\right) = 0, \qquad j \neq k$$

If we denote by $\boldsymbol{\varphi}$ the vector with components $\varphi_1, \varphi_2, \ldots, \varphi_{n-1}$, the last equation can be written

$$B \frac{\partial \boldsymbol{\varphi}}{\partial t} + G \cdot B \frac{\partial \boldsymbol{\varphi}}{\partial x} = 0$$

or

$$\frac{\partial \boldsymbol{\varphi}}{\partial t} + B^{-1}GB \frac{\partial \boldsymbol{\varphi}}{\partial x} = 0 \qquad (2.2.2\text{-}5)$$

Calling B the matrix with elements b_{js} and G the diagonal matrix the elements of which are the eigenvalues g_j ($j \neq k$), then the vector $\boldsymbol{\varphi}(\mathbf{z})$ is a solution of a linear homogeneous equation of hyperbolic type; the characteristics have slopes $g_1, g_2, \ldots, g_n$ (but not g_k); the straight line C, which forms region (R), and which has the slope g_k, is therefore not a characteristic. Consequently, there exists one and only one solution of (2.2.2-5) which takes the given values on the straight line C. Since the solution $\mathbf{z}(t, x)$ is constant on one side of C, the functions $\varphi_s(\mathbf{z})$ are constant on the straight line C; it follows that they are constant in the whole region (R), which proves the theorem.

2.2.3. Simple Waves in MFD

It is interesting to apply the preceding results to the case of MFD. Such an application will be particularly useful for solving the piston problem, i.e., for finding the motion of an electrically conducting fluid acted on by a moving boundary.

The vector $\mathbf{z}$ has components $H_x, H_y, H_z, V_x, V_y, V_z, \rho, S$. For motions depending only on t and x, the equations of MFD, Eq. (2.1.1-1) can be written in the form

$$\frac{\partial \mathbf{z}}{\partial t} + A(\mathbf{z}) \frac{\partial \mathbf{z}}{\partial x} = 0 \qquad (2.2.3\text{-}1)$$

where A is the matrix

$$A = \begin{pmatrix} 0 & 0 & 0 & 0 & 0 & 0 & 0 & 0 \\ -V_y & V_x & 0 & H_y & -H_x & 0 & 0 & 0 \\ -V_z & 0 & V_x & H_z & 0 & -H_x & 0 & 0 \\ 0 & \mu H_y/\rho & \mu H_z/\rho & V_x & 0 & 0 & \rho^{-1}\,\partial p/\partial\rho & \rho^{-1}\,\partial p/\partial S \\ 0 & -\mu H_x/\rho & 0 & 0 & V_x & 0 & 0 & 0 \\ 0 & 0 & -\mu H_x/\rho & 0 & 0 & V_x & 0 & 0 \\ 0 & 0 & 0 & \rho & 0 & 0 & V_x & 0 \\ 0 & 0 & 0 & 0 & 0 & 0 & 0 & V_x \end{pmatrix}$$

The matrix A is, with a slight change in notation, the one which was introduced in Section 2.1.1. In fact, the component H_x of $\mathbf{z}$ is not an unknown of the problem, for we have†

$$\partial H_x/\partial t = 0, \qquad H_x(t, x) = H_x(0, x)$$

The vector $\mathbf{z}$ has seven components, and we can cancel out the first column and first row of A. To solve for the simple-wave motion, we must know the eigenvalues and eigenvectors of the matrix A; the computation has already been carried out. The matrix A has the following seven eigenvalues:

$$g_1 = V_x, \qquad g_2 = V_x + U_A, \qquad g_3 = V_x - U_A$$
$$g_4 = V_x + U_-, \qquad g_5 = V_x - U_-, \qquad g_6 = V_x + U_+, \qquad g_7 = V_x - U$$

Now,

$$U_A^2 = \frac{\mu H_x^2}{\rho}$$

$$2U_\pm^2 = \frac{\partial p}{\partial \rho} + \frac{\mu H^2}{2} \pm \left\{\left(\frac{\partial p}{\partial \rho} + \frac{\mu H^2}{\rho}\right)^2 - 4\,\frac{\partial p}{\partial \rho}\,\frac{\mu H_x^2}{\rho}\right\}^{1/2}$$

The eigenvectors to the right—vectors having seven components, H_y, H_z, V_x, V_y, V_z, ρ, S—are

$$\mathbf{r}_1 = \begin{vmatrix} 0 \\ 0 \\ 0 \\ 0 \\ 0 \\ -\partial p/\partial S \\ \partial p/\partial\rho \end{vmatrix}, \qquad \mathbf{r}_{2,3} = \begin{vmatrix} \pm(\mu\rho)^{1/2} H_z \\ \mp(\mu\rho)^{1/2} H_y \\ 0 \\ -\mu H_z \\ \mu H_y \\ 0 \\ 0 \end{vmatrix}$$

† We will take $H_x(0, x) = \text{const.}$

$$\mathbf{r}_{6,7} = \begin{vmatrix} H_y U_+^2/(\rho U_+^2 - \mu H_x^2) \\ H_z U_+^2/(\rho U_+^2 - \mu H_x^2) \\ \pm U_+/\rho \\ \mp \mu H_x H_y U_+/(\rho U_+^2 - \mu H_x^2) \\ \mp \mu H_x H_z U_+/(\rho U_+^2 - \mu H_x^2) \\ 1 \\ 0 \end{vmatrix}$$

The vectors $\mathbf{r}_4$ and $\mathbf{r}_5$ can be deduced from $\mathbf{r}_6$ and $\mathbf{r}_7$ respectively by changing U_+ into U_-. The simple waves associated with the value g_1 correspond to contact discontinuities; we will not consider this case. Of the remaining cases, three are progressive waves (those corresponding to eigenvalues g_2, g_4, g_6) and three are regressive waves (those corresponding to the eigenvalues g_3, g_5, g_7). We confine attention to the first three cases; the results can easily be generalized to the other cases by appropriate changes in sign.

The simple waves associated with the eigenvalue g_2 (or g_3) are called simple MFD waves or simple Alfvén waves. The flow is completely determined when we know the function $\mathbf{z}(X)$, a solution of the system (2.2.1-2), and $X(t, x)$ defined by the relation (2.2.1-4). The system (2.2.1-2) is written

$$\frac{dH_y}{(\mu\rho)^{1/2} H_z} = \frac{dH_z}{-(\mu\rho)^{1/2} H_y} = \frac{dV_x}{0} = \frac{dV_y}{-\mu H_z} = \frac{dV_z}{\mu H_y} = \frac{d\rho}{0} = \frac{dS}{0}$$

This system has six first integrals which are solutions of Eq. (2.2.2-1) defining the Riemann invariants:

$$\varphi_1 = H_y^2 + H_z^2, \qquad \varphi_2 = V_x, \qquad \varphi_3 = \rho$$

$$\varphi_4 = S, \qquad \varphi_5 = (\mu^{-1}\rho)^{1/2} V_y + H_y, \qquad \varphi_6 = (\mu^{-1}\rho)^{1/2} V_z + H_z$$

These six quantities are the constants of the flow; they make it possible to express the (nonconstant) components of $\mathbf{z}$ as a function of one of them (nonconstant). In a simple Alfvén-wave flow, the eigenvalue $g_2(\mathbf{z})$ is constant; it follows that the straight-line characteristics are parallel. This property is true whenever the gradient of the eigenvalue $g_i(\mathbf{z})$ (computed in the Euclidean space where $\mathbf{z}$ is defined) is orthogonal to the corresponding eigenvector $\mathbf{r}_i(\mathbf{z})$:

$$\mathbf{r}_i(\mathbf{z}) \cdot \nabla g_i(\mathbf{z}) = 0$$

because in this case $g_i(\mathbf{z})$ is a Riemann invariant, or also a first integral of Eq. (2.2.1-2).

2.2.4. Study of Magnetoacoustic Simple Waves

We will deal with the two cases of fast magnetoacoustic simple waves and slow magnetoacoustic simple waves at the same time, writing U instead of U_+ or U_-. The system (2.2.1-2) which determine the function $\mathbf{z}(X)$ becomes

$$\frac{dH_y}{H_y U^2/\Delta} = \frac{dH_z}{H_z U^2/\Delta} = \frac{dV_x}{U/\rho} = \frac{dV_y}{-\mu H_x H_y U/\rho\Delta}$$

$$= \frac{dV_z}{-\mu H_x H_z U/\rho\Delta} = \frac{d\rho}{1} = \frac{dS}{0} \qquad (2.2.4\text{-}1)$$

where

$$\Delta = \rho U^2 - \mu H_x^2$$

We can use $\rho(t, x)$ as the function $X(t, x)$. We will show that $\mathbf{z}(\rho)$, i.e., in fact, H_y, H_z, V_x, V_y, V_z, can be found as a function of ρ by quadratures.

The system (2.2.4-1) has the obvious first integral S; the function $\varphi = S$ is therefore a Riemann invariant. We will construct five other first integrals of (2.2.4-1), i.e., five other Riemann invariants.

1. $\varphi_1(\mathbf{z}) = H_y/H_z$ is a first integral of the system (2.2.4-1), hence a Riemann invariant. In a simple magnetoacoustic wave, we have

$$H_y/H_z = K_1, \qquad K_1 = \text{const}$$

2. $\varphi_2(\mathbf{z}) = V_y - (H_y/H_z)\, V_z$ is also a Riemann invariant. In a simple magnetoacoustic wave flow, we have

$$V_y = K_1 V_z + K_2, \qquad K_2 = \text{const}$$

3. To construct a third Riemann invariant, we write $a^2 = \partial p/\partial\rho$ and introduce the two parameters

$$q = U^2/a^2, \qquad s = a^2/U_A{}^2 = a^2\rho/\mu H_x^2$$

The square H^2 of the magnitude of the magnetic field can be expressed as a function of q and s. In fact, the velocities U_- and U_+ of magnetoacoustic waves are solutions of the equation

$$U^4 - \left(a^2 + \frac{\mu H^2}{\rho}\right) U^2 + a^2 \frac{\mu H_x^2}{\rho} = 0$$

Replacing a^2 by $\mu H_x{}^2/\rho$ and U^2 by $(\mu H_x{}^2/\rho)qs$, we get

$$H^2 = (q-1)[s-(1/q)]\,H_x{}^2 + H_x{}^2$$
$$dH^2 = H_x{}^2\{(q-1)\,ds + [s-(1/q^2)]\,dq\}$$

It is also possible to obtain an expression for the differential dH^2 from the system (2.2.4-1). Combining the equations, we get

$$\frac{dH^2}{2\mu U^2(H^2 - H_x{}^2)/\Delta} = \frac{d\rho}{1}$$

Since

$$U^2 = (\mu H_x{}^2/\rho)\,qs$$
$$H^2 - H_x{}^2 = H_x{}^2(q-1)[s-(1/q)]$$
$$\Delta = \rho U^2 - \mu H_x{}^2 = \mu H_x{}^2(qs-1)$$

we deduce that

$$dH^2 = H_x{}^2\,2qs(q-1)\,d\rho/\rho$$

Assuming from now on that the gas is polytropic, we have

$$a^2 = f(S)\,\gamma\rho^{\gamma-1}, \qquad s = a^2/U_{\mathrm{A}}{}^2 = [\gamma f(S)/\mu H_x{}^2]\,\rho^\gamma$$

Since S and H_x are constant in the flow, we deduce that $(\gamma\,d\rho)/\rho = ds/s$. The two expressions for dH^2 yield the differential equation

$$\frac{ds}{dq} - \frac{\theta s}{q-1} + \frac{\theta}{q^2(q-1)} = 0, \qquad \theta = \frac{\gamma}{2-\gamma}$$
$$\frac{s}{(q-1)^\theta} + \theta\int^q \frac{dq}{q^2(q-1)^{\theta+1}} = K_3 \tag{2.2.4-2}$$

K_3 being a constant. Hence, the function

$$\varphi_3(\mathbf{z}) = \frac{s}{(q-1)^\theta} + \theta\int^q \frac{dq}{q^2(q-1)^{\theta+1}} \tag{2.2.4-3}$$

is a Riemann invariant which we call the magnetic Riemann invariant.

The point $q = 1$, $s = 1$ is a singular point (nodal point) for the differential equation (2.2.4-2); in the neighborhood of this point, we have

$$s - 1 = -[\gamma/(\gamma-1)](q-1) + O(q-1)^2$$

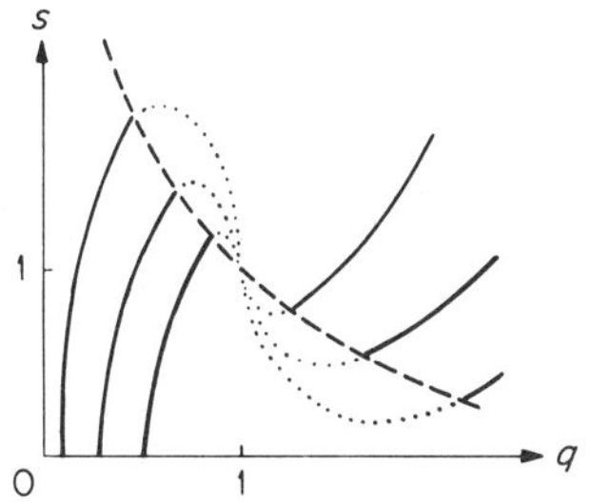

FIG. 2.2.4.a. Variation of $s(q)$.

The integral curves are shown in Fig. 2.2.4.a; on these curves, the parts shown by a broken line have no physical meaning. In fact, the velocities of propagation of magnetoacoustic waves satisfy the inequalities

$$U_- \leqslant a \leqslant U_+ , \qquad U_- \leqslant U_A \leqslant U_+$$

or

$$q \leqslant 1 \quad \text{and} \quad qs \leqslant 1 \quad \text{(slow waves)}$$

$$q \geqslant 1 \quad \text{and} \quad qs \geqslant 1 \quad \text{(fast waves)}$$

These corresponding curve segments are represented by continuous lines; on these curves, we have $ds/dq > 0$ if $\gamma \neq 2$. The magnetic Riemann invariant can be written in explicit form for certain values of the adiabatic index γ,

$$\gamma = 1 \qquad \varphi_3(\mathbf{z}) = \frac{s-1}{q-1} - 2 \log\left(1 - \frac{1}{q}\right)$$

$$\gamma = \frac{5}{3} \qquad \varphi_3(\mathbf{z}) = \frac{s-1}{(q-1)^5} + \frac{5}{2}\frac{1}{(q-1)^4} - \frac{5}{(q-1)^3} + \frac{10}{(q-1)^2}$$

$$- \frac{25}{q-1} - \frac{5}{q} - 30 \log\left(1 - \frac{1}{q}\right)$$

$$\gamma = 2 \qquad \varphi_3(\mathbf{y}) = q, \qquad 1 - sq^2 = 0$$

The integral curves corresponding to the last case are shown in Fig. 2.2.4.b.

4. Since H_x and S are constants of the flow, the parameter s is a function of ρ, and q is a function of U and ρ. We can therefore write

$$\varphi_3(\mathbf{z}) = \varphi_3(U, \rho) = K_3 , \qquad U = U(\rho, K_3)$$

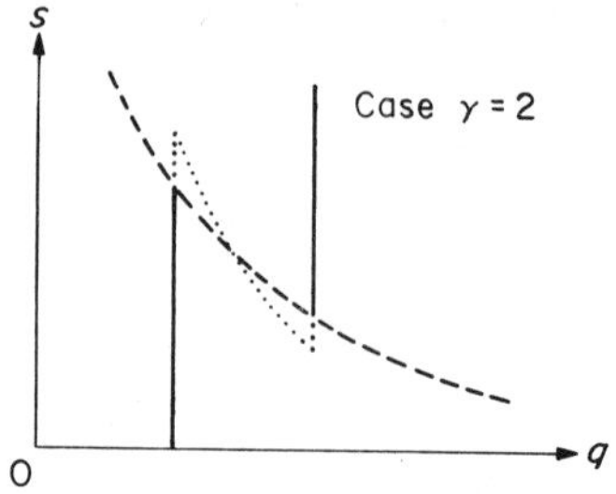

FIG. 2.2.4.b. Variation of $s(q)$.

We get from Eq. (2.2.4-1)

$$dV_x = U(\rho, K_3)\, d\rho/\rho$$

so that the function

$$\varphi_4(\mathbf{z}) = V_x - \int^{\rho} U(\rho, K_3)\, d\rho/\rho \tag{2.2.4-4}$$

constitutes a new Riemann invariant, it is a generalization of the Riemann invariant of classical aerodynamics. The integration must then be carried out assuming that K_3 is constant, and, once completed, K_3 must be replaced by $\varphi_3(\mathbf{z})$.

5. The last Riemann invariant can be determined in the same way using the preceding invariants to eliminate certain variables. We can, for instance, express ρ as a function of V_x, then U as a function of V_x; hence, q and s could be expressed as functions of V_x, and it follows that H^2, and hence H_y, could also be expressed as function of V_x. Comparing the third and the fourth ratios of Eq. (2.2.4-1), we get the differential equation

$$dV_y = -\frac{\mu H_x H_y}{\rho U^2 - \mu H_x^2}\, dV_x$$

so that the function

$$\varphi_5(\mathbf{z}) = V_y + \int^{V_x} \frac{\mu H_x H_y}{\rho U^2 - \mu H_x^2}\, dV_x \tag{2.2.4-5}$$

is the fifth Riemann invariant (in addition to the invariant S).

The existence of Riemann invariants, the number of which is equal to the number of unknowns minus one, enables us to reduce the integration of Eq. (2.2.4-1) to quadratures. The quantities H_x and S are constant, and we will express the quantities H_z, V_z, and V_x as a function of the density ρ regarded as variable; we will then have

$$H_y = K_1 H_z \qquad \text{and} \qquad V_y = K_1 V_z + K_2$$

Equation (2.2.4-1) can be written

$$\frac{dV_z}{-\mu H_x H_z U/\rho\Delta} = \frac{dV_x}{U/\rho} = d\rho$$

If we notice that

$$U^2 = \frac{\mu H_x^2}{\rho} q\, s, \qquad \Delta = \mu H_x^2(qs - 1)$$

$$H_y^2 + H_z^2 = (1 + K_1^2)\, H_z^2 = H_x^2(q - 1)(s - 1/q)$$

we get

$$\begin{aligned} \frac{dV_z}{d\rho} &= -\frac{1}{\rho} \frac{\mu^{1/2} H_x}{\rho^{1/2}(1 + K_1^2)^{1/2}} \left\{ \frac{q - 1}{q - (1/s)} \right\}^{1/2} \\ \frac{dV_x}{d\rho} &= \frac{1}{\rho} \frac{\mu^{1/2} H_x}{\rho^{1/2}} (qs)^{1/2} \end{aligned} \tag{2.2.4-6}$$

Since q is a known function of s and s a known function of ρ,

$$s = \gamma[f(S)/\mu H_x^2]\, \rho^\gamma$$

The integration of the system (2.2.4-1) which determines the vector $\mathbf{z}(\rho)$ for simple magnetoacoustic waves is reduced to quadratures.

CHAPTER 3

PROPAGATION OF SHOCK WAVES

Simple waves, such as those studied in the previous chapter, can, as time passes, develop into strong discontinuities or shock waves. The appearance of shock waves is a fundamental phenomenon of gas dynamics, and the object of this chapter is to study these in detail. In Section 3.1 we will study the general properties of shocks, both thermodynamic and geometrical, and also the actual solution of the shock equations. Section 3.2 deals with stability; stable shocks, the only kind realized in practice, are the fast and slow shocks in the case of nonreacting media; in the case of a reacting media, there exist six types of stable shock: four detonation waves and two weak deflagration waves. Section 3.3 is devoted to the study of wedge and piston problems.

3.1. General Properties

3.1.1. Geometrical Properties

We consider the propagation of shock waves in a fluid which is compressible with negligible thermal conductivity and electrical resistivity. The shock equations, established in the first chapter, express the conservation of the following quantities across a shock:

$$(V_n - U)\mathbf{H} - H_n\mathbf{V}, \qquad (V_n - U)\rho\mathbf{V} + \boldsymbol{\pi}, \qquad (V_n - U)\rho$$
$$(V_n - U)\,\rho\{\epsilon_i + \tfrac{1}{2}V^2 + (\mu H^2/2\rho)\} + \boldsymbol{\pi}\cdot\mathbf{V} \qquad (3.1.1\text{-}1)$$

with

$$\pi = (p + \tfrac{1}{2}\mu H^2)\mathbf{n} - \mu H_n \mathbf{H}$$

$$\epsilon_i = \epsilon_i(p, \rho)$$

In the particular case of a polytropic gas, which we frequently consider, we have

$$\epsilon_i(p, \rho) = \frac{1}{\gamma - 1}\frac{p}{\rho} = \epsilon_v T$$

The shock equations permit us to determine the flow conditions behind the shock when the conditions ahead and the shock velocity are known; the velocity of the shock is defined by the vector **n** and the scalar U. The solution will be given in Section 3.1.5, but first we shall establish some properties.

The components of the magnetic field and the velocity in the plane tangential to the shock wave are denoted by $\mathbf{h} = \mathbf{n} \times (\mathbf{H} \times \mathbf{n})$ and by $\mathbf{v} = \mathbf{n} \times (\mathbf{V} \times \mathbf{n})$. The symbol $[Q]$ indicates the difference $Q_2 - Q_1$ of the values of Q on the two sides of the shock wave. The first equation of (3.1.1-1) resolves into the following two equations:

$$\begin{aligned} -U[H_n] &= 0 \\ [(V_n - U)\mathbf{h}] - H_n[\mathbf{v}] &= 0 \end{aligned} \qquad (3.1.1\text{-}2)$$

Since there is always a frame of reference in which U is nonzero, we have $[H_n] = 0$. In a similar fashion, the second equation of (3.1.1-1) resolves into the pair

$$\rho(V_n - U)[V_n] + [p + \tfrac{1}{2}\mu h^2] = 0 \qquad (3.1.1\text{-}3a)$$

or

$$[\rho(V_n - U)^2] + [p + \tfrac{1}{2}\mu h^2] = 0 \qquad (3.1.1\text{-}3b)$$

and

$$\rho(V_n - U)[\mathbf{v}] - H_n[\mathbf{h}] = 0 \qquad (3.1.1\text{-}3c)$$

The last equation of (3.1.1-1) can then be replaced by the following:

$$\rho(V_n - U)[\epsilon_i + \tfrac{1}{2}(\mathbf{V} - \mathbf{n}U)^2 + (\mu H^2/2\rho)] + [\pi \cdot (\mathbf{V} - \mathbf{n}U)] = 0 \quad (3.1.1\text{-}4)$$

The second equation of (3.1.1-2) can be multiplied by

$\rho(V_n - U) = \rho_1(V_{n_1} - U) = \rho_2(V_{n_2} - U)$, and Eq. (3.1.1-3.c) by $H_n = H_{n_1} = H_{n_2}$; adding the two yields

$$[\{\rho(V_n - U)^2 - \mu H_n^2\}\mathbf{h}] = 0 \qquad (3.1.1\text{-}5)$$

We will write $\rho(V_n - U)^2 - \mu H_n^2 = A$.

Equation (3.1.1-5), which can replace two of the scalar shock equations, simplifies in three cases, which it is convenient to examine before studying the general case. The three special cases are as follows: the speed of propagation of the shock is equal to the Alfvén speed before the shock ($A_1 = 0$); the shock speed is equal to the Alfvén speed after the shock ($A_2 = 0$); the two Alfvén speeds before and after the shock are both equal to the speed of propagation of the shock ($A_1 = A_2 = 0$). Let us look at each of these particular cases.

(i) *Switch-Off Shocks.* $A_1 = 0$, $A_2 \neq 0$. The shock speed is equal to the Alfvén speed ahead of the shock. Equation (3.1.1-5) gives $\mathbf{h}_2 = 0$, even if $\mathbf{h}_1 \neq 0$; this means that the shock destroys the tangential magnetic field $\mathbf{h}$ and also the normal electric field $\mathbf{n}\, E_n = -\mathbf{v} \times \mu\mathbf{h}$.

(ii) *Switch-On Shocks.* $A_1 \neq 0$, $A_2 = 0$. The shock speed equals the Alfvén speed behind the shock. This situation can exist only if $\mathbf{h}_1 = 0$ (otherwise, $\mathbf{h}_2$ will be infinite). Equation (3.1.1-5) then shows that we can have $\mathbf{h}_2 \neq 0$, which means that the shock creates a tangential magnetic field and a normal electric field.

(iii) *Alfvén Shocks.* $A_1 = 0$, $A_2 = 0$. The shock speed equals the Alfvén speed both ahead of and behind the shock. Equation (3.1.1-5) is always satisfied, but we have

$$V_{n_1} - U = \frac{\mu H_n^2}{\rho(V_n - U)}, \qquad V_{n_2} - U = \frac{\mu H_n^2}{\rho(V_n - U)}$$

We deduce that $[V_n - U] = 0$, while Eq. (3.1.1-4) is always satisfied. The number of shock equations is reduced by one, but this is balanced by the fact that the shock speed is determined and no longer arbitrary. We have

$$V_n - U = \pm\mu H_n/(\mu\rho)^{1/2}$$

The shock equations are then reduced to the following:

$$[H_n] = 0, \qquad [\rho] = 0, \qquad [p] = 0, \qquad [\mathbf{h}^2] = 0, \qquad [\mathbf{v}] = \mu[\mathbf{h}]/(\mu\rho)^{1/2} \qquad (3.1.1\text{-}6)$$

The corresponding shocks are called Alfvén shocks. In these, the tangential magnetic field **h** is conserved in magnitude but is rotated through an arbitrary angle; in practice, this angle is determined by the conditions behind the shock.

Apart from the three particular cases just examined, we can establish some geometrical properties satisfied by discontinuities across a shock wave. These properties, except for the first, are not valid for switch-on shocks, switch-off shocks, and Alfvén shocks.

Property 1. The discontinuity of the magnetic field is tangential.

Property 2. The direction of the tangential magnetic field is conserved across the shock.

Property 3. The plane of the vectors **n** and **H** is invariant across a shock.

Property 4. If the magnetic field ahead of the shock is zero, it will also be zero after the shock.

Property 5. The discontinuity of the tangential velocity is collinear with the discontinuity of the magnetic field.

Property 6. The velocity component orthogonal to the plane (**n**, **H**) is continuous across the shock.

The following properties are related to the electrical field. Since the electrical resistivity is negligible, we can write successively:

$$\begin{aligned}
\mathbf{E} &= -\mathbf{V} \times \mathbf{H} = \mathbf{n}\, E_\mathrm{n} + \mathbf{e} \\
\mathbf{n}E_\mathrm{n} &= -\mathbf{v} \times \mu\mathbf{h} \\
\mathbf{n}AE_\mathrm{n} &= -\mathbf{v} \times \mu A\mathbf{h} \\
\mathbf{n}[AE_\mathrm{n}] &= -[\mathbf{v}] \times \mu A\mathbf{h} = 0 \\
\mathbf{e} &= -\mathbf{v} \times \mathbf{n}\mu H_\mathrm{n} - \mathbf{n}V_\mathrm{n} \times \mu\mathbf{h} \\
\mathbf{e} &= -\mathbf{n} \times \{V_\mathrm{n}\mu\mathbf{h} - \mu H_\mathrm{n}\mathbf{v}\} \\
[\mathbf{e}] &= -\mathbf{n} \times \mu[V_\mathrm{n}\mathbf{h} - H_\mathrm{n}\mathbf{v}] \\
[\mathbf{e}] &= -\mathbf{n} \times \mu U[\mathbf{h}]
\end{aligned}$$

Property 7. If the normal electrical field is zero ahead of the shock, it will also be zero after the shock.

Property 8. In a coordinate system fixed in the shock, the tangential electrical field is continuous across the shock.

Property 9. In a coordinate system fixed in the shock, if the electrical field ahead of the shock is zero, it will also be zero behind it.

3.1.2. The Shock Surfaces

Before solving the shock equations (this will be done in Section 3.1.5), it is interesting to give a graphical representation of the shock equations. Since the (**n**, **H**) plane is invariant across the shock, we denote the components of the velocity vector **V** in the three directions **n**, (**n** × **H**) × **n**, and **n** × **H** by V_n, v, and w. The components of the magnetic field in the same direction are H_n, h, and 0. Through the shock, H_n and w remain invariant, the quantities h, $V_n - U$, v, p, and ρ vary, but from the shock equation, the points in a five-dimensional space which have as coordinates the values of these quantities ahead of and behind the shock are located on the surfaces

$$(V_n - U)h - H_n v = E, \qquad E = \text{const}$$

$$(V_n - U)\rho = G, \qquad G = \text{const}$$

$$G^2\tau + p + \frac{\mu h^2}{2} = F_n, \qquad F_n = \text{const}$$

$$Gv - \mu H_n h = \text{const}$$

$$G\left\{\epsilon_i + \frac{(V_n - U)^2 + v^2}{2} + \frac{\mu H^2}{2\rho}\right\} + (V_n - U)\left\{p + \frac{\mu H^2}{2}\right\}$$
$$- \mu H_n\{vh + (V_n - U)H_n\} = \text{const}$$

$$\tau = \frac{1}{\rho}$$

Eliminating the velocity components $V_n - U$ and v, we find

$$G^2\tau + p + \frac{\mu h^2}{2} = F_n, \qquad h\left(\tau - \frac{\mu H_n^2}{G^2}\right) = F_t$$
$$\epsilon_i + p\tau + \frac{G^2\tau^2}{2} + \frac{G^2\tau^2 h^2}{2H_n^2} = K \tag{3.1.2-1}$$

where F_t and K are constants.

In a three-dimensional space, h, τ, RT, the points representing the states which can be connected across the shock lie on the following surfaces:

$$(S_1) \qquad \epsilon_i + p\tau = K - \tfrac{1}{2}G^2\tau^2 - (\mu h^2\tau^2/2\tau^*)$$
$$(S_2) \qquad p\tau = (F_n - \tfrac{1}{2}\mu h^2)\tau - G^2\tau^2$$
$$(S_3) \qquad h(\tau - \tau^*) = F_t$$

where we have put $\tau^* = \mu H_n^2/G^2$.

The fluid states which can be connected across a shock are represented by the points of intersection of these three surfaces which are located in the region $\tau > 0$, $RT > 0$.

In the case of a polytropic perfect gas, we have

$$p\tau = RT, \qquad \epsilon_i = \epsilon_v T = RT/(\gamma - 1)$$

The surfaces (S_1), (S_2), and (S_3), which are therefore all algebraic, are represented in Fig. 3.1.2.a. The sections $h =$ const and $\tau =$ const of the surfaces (S_1) or (S_2) are parabolas. The surfaces (S_1) and (S_2) intersect along a curve (c) and its projection on the h, τ plane has the equation

$$\mu h^2 = \frac{[(\gamma + 1)/(\gamma - 1)]\, G^2\tau^2 - [2\gamma/(\gamma - 1)]\, F_n\tau + 2K}{(\tau/\tau^*)\{\tau - [\gamma/(\gamma - 1)]\, \tau^*\}} \qquad (3.1.2\text{-}2)$$

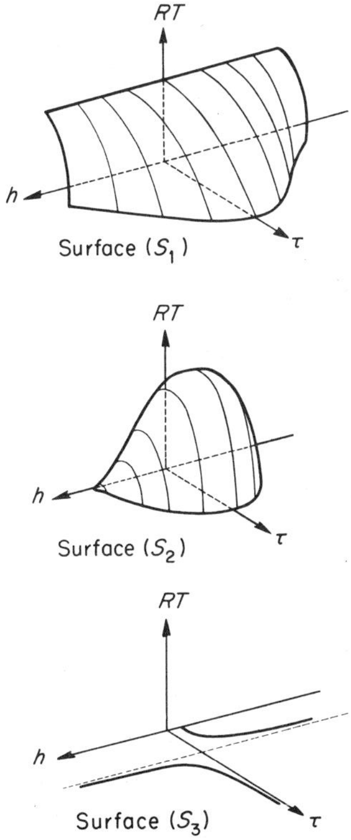

FIG. 3.1.2.a. Representation of the algebraic surfaces (S_1), (S_2), and (S_3).

The curve (c) and the cylinder (S_3) have at most four common points. Two cases are represented in Fig. 3.1.2.b, in each of which we find four real points in the region $\tau > 0$, $RT > 0$. These points are num-

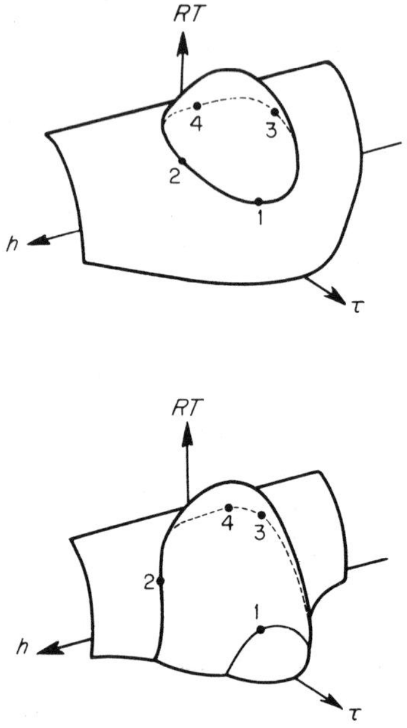

FIG. 3.1.2.b. Intersection (c) of two surfaces (S_1) and (S_2).

bered from 1 to 4 following decreasing values of τ $(\tau_1 \geqslant \tau_2 \geqslant \tau_3 \geqslant \tau_4)$; we will prove in Section 3.1.4 that they are also numbered in the same way following increasing values of the specific entropy:

$$s_1 \leqslant s_2 \leqslant s_3 \leqslant s_4$$

By the second law of thermodynamics, the specific entropy can not decrease; the possible shocks, totalling six, are then those joining a point i to a point j $(j > i)$. The shock speed is $V_n - U = G\tau$; the shock $1 \to 2$ is called a fast shock, the shock $3 \to 4$ is called a slow shock, and the other shocks $i \to j$ with $(i = 1, 2)$ and $(j = 3, 4)$ are called intermediate shocks.

3.1.3. The Hugoniot Relation

Given that H_n and $\rho(V_n - U) = G$ stay constant across a shock, the shock equations obtained from Faraday's law and the kinematic resultant can be written in the form

$$\begin{aligned} G[\tau h] - H_n[v] &= 0 \\ G[v] - \mu H_n[h] &= 0 \\ G^2[\tau] + [p + \tfrac{1}{2}\mu h^2] &= 0 \end{aligned} \tag{3.1.3-1}$$

The equation of conservation of energy

$$G[\epsilon_i + \tfrac{1}{2}(G^2\tau^2 + v^2) + \tfrac{1}{2}\tau\mu H^2] + G[\tau(p + \tfrac{1}{2}\mu H^2)] - \mu H_n[vh + G\tau H_n] = 0 \tag{3.1.3-2}$$

can be simplified, using relations (3.1.3-1). It is first written in the form

$$[\epsilon_i] + [\tau p] + \tfrac{1}{2}G^2[\tau^2] + \mu[\tau h^2] + [\tfrac{1}{2}v^2 - (\mu H_n/G)\, vh] = 0$$

If we use the relations

$$\frac{v^2}{2} - \frac{\mu H_n}{G} vh = \frac{1}{2}\left(v - \frac{\mu H_n}{G} h\right)^2 - \frac{\mu^2 H_n^2}{2G^2} h^2$$

$$[Q^2] = [Q](Q_1 + Q_2)$$

we can write

$$\left[\frac{v^2}{2} - \frac{\mu H_n}{G} vh\right] = -\frac{\mu H_n^2}{2G^2}[h](h_1 + h_2)$$

Next, using the relations

$$[\tau p] = \tfrac{1}{2}(p_1 + p_2)[\tau] + \tfrac{1}{2}(\tau_1 + \tau_2)[p]$$

$$\tfrac{1}{2}G^2[\tau^2] = \tfrac{1}{2}G^2(\tau_1 + \tau_2)[\tau] = -\tfrac{1}{2}(\tau_1 + \tau_2)[p + \tfrac{1}{2}\mu h^2]$$

$$G^2[\tau h] - \mu H_n^2[h] = 0$$

we can write Eq. (3.1.3-2) in the following form, called Hugoniot's equation:

$$[\epsilon_i] + \tfrac{1}{2}[\tau](p_1 + p_2) + \tfrac{1}{4}[\tau]\,\mu[h]^2 = 0 \tag{3.1.3-3}$$

Hugoniot's equation only contains the variables τ and p, since we have

$$[h] = h_2 - h_1 \qquad \text{and} \qquad h_2 = \frac{\tau_1 - \tau^*}{\tau_2 - \tau^*} h_1$$

where

$$\tau^* = \mu H_n^2/\rho^2(V_n - U)^2$$

The Hugoniot equation forms, together with Eq. (3.1.3-1), a complete system of shock equations. With the shock equations we can calculate $\mathbf{H}_2$, $\mathbf{V}_2$, p_2, and τ_2 once $\mathbf{H}_1$, $\mathbf{V}_1$, p_1, and τ_1 and the vector $\mathbf{n}U$ are known. In other words, given a state (1) among the states that could result from a shock, the vectors $\mathbf{H}$, $\mathbf{V}$ and the scalars p, τ depend only on the vector $\mathbf{n}U$.

With the Hugoniot equation we associate the function $\mathscr{H}(p, \tau)$, called the Hugoniot function, and defined by the formula

$$\mathscr{H}(p, \tau) \equiv \epsilon_i(p, \tau) - \epsilon_i(p_1, \tau_1) + \frac{\tau - \tau_1}{2}\left\{p + p_1 + \frac{\mu(h - h_1)^2}{2}\right\}$$

or

$$\mathscr{H}(p, \tau) \equiv \epsilon_i(p, \tau) - \epsilon_i(p_1, \tau_1) + \frac{\tau - \tau_1}{2}\left\{p + p_1 + \frac{\mu h_1^2}{2}\left(\frac{\tau - \tau_1}{\tau - \tau^*}\right)^2\right\} \tag{3.1.3-4}$$

The curve $\mathscr{H}(p, \tau) = 0$, called the Hugoniot curve, represents in the p, τ plane the states that could be attained across a shock from the given state p_1, τ_1.

The Hugoniot curves corresponding to different values of the magnetic field h_1 intersect and have a common tangent at the point $\tau = \tau_1$, $p = p_1$. We know that the curve $\mathscr{H}(p, \tau) = 0$, which corresponds to the state $h_1 = 0$, has a common tangent with the isentropic curve, along which we have $d\epsilon_i + p\, d\tau = 0$, at the point $\tau = \tau_1$, $p = p_1$; the Hugoniot curve lies above the isentropic curve for $\tau < \tau_1$ and below it for $\tau > \tau_1$ (Courant & Friedrichs, 1948, p. 143).

For $h_1 \neq 0$, we have

$$\mathscr{H}(\tau, p) \equiv \mathscr{H}_0(\tau, p) + \frac{\mu h_1^2}{4} \frac{(\tau - \tau_1)^3}{(\tau - \tau^*)^2}$$

It follows that all the Hugoniot curves touch the isentropic curve at the point $\tau = \tau_1$, $p = p_1$, lie above it for $\tau < \tau_1$ and below it for $\tau > \tau_1$. These curves are shown in Fig. 3.1.3.a.

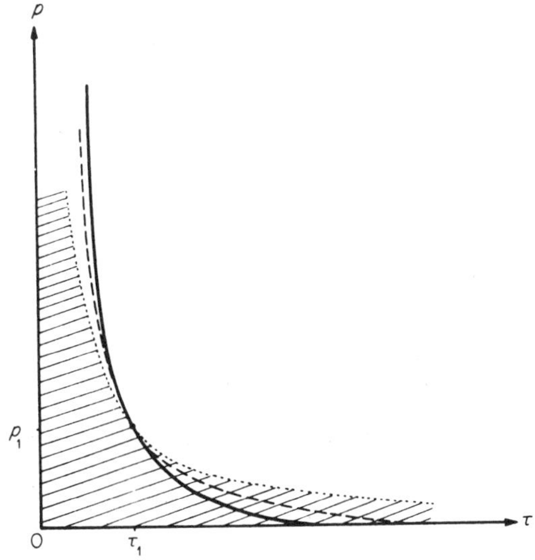

FIG. 3.1.3.a. Hugoniot curves. Dotted curve: isentropic curve; dashes: $\mathscr{H}_0(\tau, p) = 0$; solid curve: $\mathscr{H}_0(\tau, p) = 0$.

In the case of a polytropic gas, $(\gamma - 1)\,\epsilon_i(\tau, p) = \tau p$, the Hugoniot curves and the isentropic curves will have the equations

$$\frac{p}{p_1} = \frac{[\tau_1/(\gamma - 1)] - \frac{1}{2}(\tau - \tau_1)\{1 + (\mu h_1^2/2p_1)[(\tau - \tau_1)/(\tau - \tau^*)]^2\}}{[\tau/(\gamma - 1)] + [(\tau - \tau_1)/2]} \qquad (3.1.3\text{-}5)$$

$$\left(\frac{p}{p_1}\right) = \left(\frac{\tau}{\tau_1}\right)^{-\gamma} \qquad (3.1.3\text{-}6)$$

We can verify that these curves intersect and have a common tangent at the point $\tau = \tau_1$, $p = p_1$ and take the relative position shown in Fig. 3.1.3.a.

3.1.4. Thermodynamic Properties

The geometrical properties established in Section 3.1.1 are valid for any $\epsilon_i(p, \tau)$, which defines the specific internal energy. On the other hand, the properties concerning the relative positions of the isentropic and Hugoniot curves $\mathscr{H}(\tau, p) = 0$ are not satisfied unless some additional hypotheses are used (Courant & Friedrichs, 1948, p. 143); it is sufficient, for example, that the function $p = p(\tau, s)$, which represents the equation of state of the gas, satisfies the following hypotheses (those of Weyl):

$$\partial p/\partial s > 0, \qquad \partial p/\partial \tau < 0, \qquad \partial^2 p/\partial \tau^2 > 0 \tag{3.1.4-1}$$

These very general hypotheses will allow us to establish the following properties:

Theorem. Across a shock, the pressure and the specific mass increase.

In other words, shocks are compressive. This result is a consequence of the position of the Hugoniot curves relative to the isentropic curve. The isentropic curve divides the first quadrant of the plane in two regions. The points situated in the region containing the origin represent states for which the specific entropy s is smaller than the value s_1 corresponding to the state τ_1, p_1; for points situated in the other region, we have $s > s_1$. Across a shock, the entropy cannot decrease, by the second principle of thermodynamics, and hence only the portions of the curves in the region $s > s_1$ can exist physically; on these portions, we have: $p > p_1$ and $\tau < \tau_1$, which proves that the pressure and specific mass increase across a shock.

It is possible to give another proof of this result based directly on the Hugoniot equation:

$$\mathscr{H} \equiv \epsilon_i(p, \tau) - \epsilon_i(p_1, \tau_1) + \tfrac{1}{2}(\tau - \tau_1)\{p + p_1 + \tfrac{1}{2}\mu(h - h_1)^2\} = 0$$

For a shock representing a change from a state (1) to a state (2), we have $\mathscr{H}(p_2, \tau_2) = 0$ and $s_2 > s_1$. We will show that the hypothesis $\tau_2 - \tau_1 > 0$ leads to a contradiction, provided the gas satisfies the Weyl hypothesis. In fact, if we consider ϵ_i as a function of τ and s, we have

$$p = -\partial \epsilon_1/\partial \tau, \qquad \partial^2 p/\partial \tau^2 > 0$$

Hence, by a property of convex functions,

$$\epsilon_i(\tau_1, s_1) - \epsilon_i(\tau_2, s_1) = \int_1^2 p(\tau, s_1)\, d\tau < (\tau_2 - \tau_1)[p(\tau_1, s_1) + p(\tau_2, s_1)]/2$$

On the other hand, since $\tau = \partial \epsilon_i/\partial s$ is positive, we can write $\epsilon_i(\tau_2, s_2) > \epsilon_i(\tau_2, s_1)$; hence,

$$\epsilon_i(\tau_1, s_1) - \epsilon_i(\tau_2, s_2) < \epsilon_i(\tau_1, s_1) - \epsilon_i(\tau_2, s_1) < (\tau_2 - \tau_1)[p(\tau_1, s_1) + p(\tau_2, s_1)]/2$$

The hypothesis $(\partial p/\partial s) > 0$ yields $p(\tau_2, s_2) > p(\tau_2, s_1)$; hence,

$$\epsilon_i(\tau_1, s_1) - \epsilon_i(\tau_2, s_2) < (\tau_2 - \tau_1)[p(\tau_1, s_1) + p(\tau_2, s_2)]/2$$

which, with

$$\frac{\tau_2 - \tau_1}{2} \frac{\mu(h_2 - h_1)^2}{2} > 0$$

gives $\mathscr{H}(p_2, \tau_2) > 0$, contradicting the Hugoniot relation $\mathscr{H}(p_2, \tau_2) = 0$.

Across a shock, we will then have $\tau_2 - \tau_1 \leqslant 0$, which yields

$$p(\tau_1, s_1) < p(\tau_2, s_2), \qquad \partial p/\partial \tau < 0$$

$$p(\tau_2, s_1) < p(\tau_2, s_2), \qquad \partial p/\partial s > 0$$

hence, $p(\tau_2, s_2) > p(\tau_1, s_1)$; i.e., shocks are compressive.

It follows from these arguments that, in a shock, the specific mass and specific entropy change in opposite directions. The points 1, 2, 3, 4 shown in Fig. 3.1.2.b represent the states that would be related by a shock. Having assumed that $\tau_1 \geqslant \tau_2 \geqslant \tau_3 \geqslant \tau_4$, we have $s_1 \leqslant s_2 \leqslant s_3 \leqslant s_4$.

It is also interesting to note that the velocities of shock propagation at these points are related by certain inequalities with the speed of propagation of linearized waves, the Alfvén speed, and the magnetic acoustic speeds; these are given below. In fact, if we take the last of Eqs. (3.1.2-1) and (3.1.2-3) and eliminate the tangential magnetic field using Eq. (3.1.2-2), we get

$$p + G^2\tau + \frac{1}{2} \frac{\mu F_t^2}{(\tau - \tau^*)^2} = F_n \qquad (3.1.4\text{-}2)$$

$$\epsilon_i(p, \tau) + p\tau + \frac{G^2\tau^2}{2} + \frac{\mu\tau^2 F_t^2}{2\tau^*(\tau - \tau^*)^2} = K \qquad (3.1.4\text{-}3)$$

The values $\tau_1, \tau_2, \tau_3, \tau_4$ are the abscissas of the points of intersection in the plane (τ, p) of the curves defined by the previous equations. The curve defined by the first equation is called the Rayleigh curve and is represented as a full line in Fig. 3.1.4.a. This curve has an asymptote $\tau = \tau^*$ and is concave downward. We therefore consider the family of curves obtained by varying the constant K in Eq.

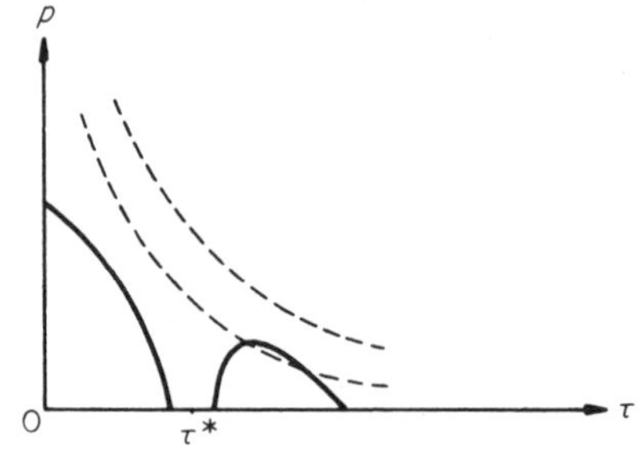

FIG. 3.1.4.a. Rayleigh curve (solid lines) and curves of constant entropy satisfying the Weyl inequalities (dashed lines).

(3.1.4-3); the constants F_t and F_n, τ^*, and G stay constant. We have, for a displacement on the Rayleigh curve,

$$dp + G^2\, d\tau = [\mu F_t^2/(\tau - \tau^*)^3]\, d\tau$$

$$d\epsilon_i + p\, d\tau + \tau\, dp + G^2 \tau\, d\tau = [\mu F_t^2 \tau/(\tau - \tau^*)^3]\, d\tau + dK$$

which gives $dK = T\, ds$.

For any displacement on the Rayleigh curve between two states related by a shock, the variation of K is zero; we then have, for example,

$$\int_1^2 T\, ds = 0$$

T being positive, the specific entropy s passes through an extreme value between the points 1 and 2. But for any gas satisfying the Weyl inequalities, the curves of constant entropy in the (τ, p) plane have a negative slope and are concave upward; these curves are shown as broken lines in Fig. 3.1.4.a; the curves for larger s are further away from the origin. It follows from these properties that there exists at most one extremum on each section of the Rayleigh curve and that this extremum is a maximum; on the other hand, we have $\tau_1 \geqslant \tau_2 > \tau^* > \tau_3 \geqslant \tau_4$. The condition $\tau_1 > \tau^*$ is then written

$$(V_{n_1} - U)^2 > \mu H_{n_1}^2/\rho_1 = U_{A_1}^2$$

which proves that at the point 1 (and also at the point 2) the shock speed is greater than the Alfvén speed; at points 3 and 4 the shock speed is less than the Alfvén speed.

We shall compare the shock speeds at the four points 1, 2, 3, and 4 with the magnetic acoustic speeds U_- and U_+ at these points and establish the following inequalities:

$$\begin{aligned} U_{+1} &< | V_n - U |_1 \\ U_{A_2} &< | V_n - U |_2 < U_{+2} \\ U_{-3} &< | V_n - U |_3 < U_{A_3} \\ &\phantom{<} | V_n - U |_4 < U_{-4} \end{aligned} \tag{3.1.4-4}$$

We denote the points of specific maximum entropy on the Rayleigh curve by M (lying between 1 and 2) and N (lying between 3 and 4). For any displacement on the Rayleigh curve, we have

$$dp = \frac{\partial p}{\partial s}\, ds + \frac{\partial p}{\partial \tau}\, d\tau = -G^2\, d\tau + \frac{\mu F_t^2}{(\tau - \tau^*)^3}\, d\tau$$

At the points M and N, ds is zero, so that

$$\frac{\partial p}{\partial \tau} \equiv -\rho^2 \frac{\partial p}{\partial \rho} = -G^2 - \frac{\mu F_t^2}{(\tau - \tau^*)^3}$$

or

$$(V_n - U)^2 \{\rho(V_n - U)^2 - \mu H^2\} - \frac{\partial p}{\partial \rho} \{\rho(V_n - U)^2 - \mu H_n^2\} = 0$$

as can be seen by replacing G by $\rho(V_n - U)$ and F_t by $h(\tau - \tau^*)$. The magnetic acoustic speeds are such that

$$f(U_\pm^2) \equiv U_\pm^2\{\rho U_\pm^2 - \mu H^2\} - \frac{\partial p}{\partial \rho} \{\rho U_\pm^2 - \mu H_n^2\} = 0$$

On the Rayleigh curve, we have

$$f\{(V_n - U)^2\} \equiv G^2\tau^2 \left\{G^2\tau - \mu H_n^2 - \frac{\mu F_t^2}{(\tau - \tau^*)^2}\right\} - \frac{\partial p}{\partial \rho} \{G^2\tau - \mu H_n^2\}$$

This is a continuous function of τ (for $\tau \neq \tau^*$) zero at the points M and N, positive for $\tau = 0$. We deduce that

$$\begin{aligned} f\{(V_n - U)^2\} &> 0 \qquad \text{at points 1 and 4} \\ &< 0 \qquad \text{at points 2 and 3} \end{aligned}$$

which, associated with the previous results, establishes the inequalities (3.1.4-4).

3.1.5. Solution of the Shock Equations

The shock equations have been written in different forms in the preceding sections; in the case of polytropic gas, all these equations are algebraic and their solution reduces to the solution of one cubic equation.

To form this equation, we can use Eq. (3.1.2-2) and substitute $F_t/(\tau - \tau^*)$ for the tangential magnetic field h. We then get the following equation of the fourth order in τ:

$$\left(\frac{\gamma+1}{\gamma-1} G^2\tau^2 - \frac{2\gamma}{\gamma-1} F_n\tau + 2K\right)(\tau-\tau^*)^2 - \frac{\mu F_n^{\,2}}{\tau^*}\tau\left(\tau - \frac{\gamma}{\gamma-1}\tau^*\right) = 0 \tag{3.1.5-1}$$

where

$$G = \frac{V_{n_1} - U}{\tau_1}, \qquad \tau^* = \frac{\mu H_n^{\,2}}{G^2}$$

$$F_n = G^2\tau_1 + p_1 + \frac{\mu h_1^{\,2}}{2}, \qquad F_t = h_1(\tau_1 - \tau^*)$$

$$K = \frac{\gamma}{\gamma-1} p_1\tau_1 + \frac{G^2\tau_1^{\,2}}{2} + \frac{\mu h_1^{\,2}\tau_1^{\,2}}{2\tau^*}$$

Equation (3.1.5-1) gives the value of $\rho_2 = 1/\tau_2$ behind the shock. We note that this equation admits the root $\tau = \tau_1$; the other three roots determine (unless $\tau < \tau_1$) the possible values of τ_2. Once τ_2 is found, we get

$$\begin{aligned} V_{n_2} &= G\tau_2 + U, \qquad h_2 = h_1\frac{\tau_1 - \tau^*}{\tau_2 - \tau^*} \\ p_2 &= p_1 + G^2(\tau_1 - \tau_2) + \frac{h_1^{\,2} - h_2^{\,2}}{2} \\ v_2 &= v_1 + \frac{\mu H_n}{G}(h_2 - h_1) \end{aligned} \tag{3.1.5-2}$$

It is possible to give a more systematic form to the fundamental cubic equation. For this, we will introduce the dimensionless numbers M^- and M^+ which are equal to the ratios of slow and fast magneto-acoustic speeds to the Alfvén speed. We will also introduce the Alfvén number

$$M_A = \frac{V_n - U}{(\mu H_n^{\,2}/\rho)^{1/2}}$$

The Alfvén number M_{A_2} behind the shock is then determined as a function of the given flow quantities ahead of the shock by the following relation of the third degree:

$$M_{A_1}^2 = \frac{\left\{ \begin{array}{l} (M_{A_2}^2 - 1)^2 \{(\gamma + 1) M_{A_2}^2 - 2M_1^{+2} M_1^{-2}\} \\ \qquad - (1 - M_1^{-2})(M_1^{+2} - 1)(\gamma M_{A_2}^2 - \gamma - 1) M_{A_2}^2 \end{array} \right\}}{(\gamma - 1)(M_{A_2}^2 - 1)^2 + (1 - M_1^{-2})(M_1^{+2} - 1)\{(2 - \gamma) M_{A_2}^2 + \gamma - 1\}} \tag{3.1.5-3}$$

Once M_{A_2}, is known, we have

$$\rho_2 = \frac{1}{M_{A_2}^2} \frac{\rho_1^2 (V_{n_1} - U)^2}{\mu H_{n_1}^2} \tag{3.1.5-4}$$

We then compute the other unknowns with Eq. (3.1.5-2).

When the values of the parameters of the flow upstream are given, except for the velocity, i.e., when M_1^- and M_1^+ are fixed but M_{A_1} is variable, the locus of the point with coordinates M_{A_1} and M_{A_2} in the M_{A_1}, M_{A_2} plane is called the shock polar, this is shown in Fig. 3.1.5.a. The segments of the polar along which the specific entropy increases are those for which $\rho_2 > \rho_1$ or $M_{A_1} < M_{A_2}$. These arcs

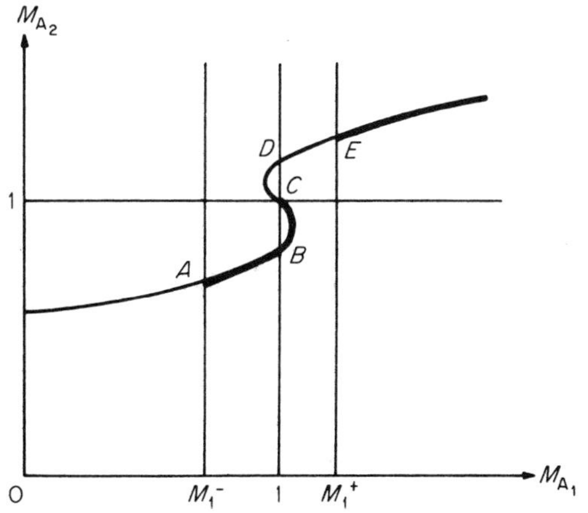

FIG. 3.1.5.a. Shock polar: $\gamma = 5/3$, $M_1^{-2} = (U_1^-/U_{A_1})^2 = 1/2$, $M_1^{+2}(U_1^+/U_{A_1})^2 = 3/2$. Thick line represents increasing specific entropy.

are shown by the thick line. On the shock polar, we have indicated the five points

A	$M_{A_1} = M_1^-$	$M_{A_2} = M_2^-$
B	$M_{A_1} = 1$	$M_{A_2} \neq 1$
C	$M_{A_1} = 1$	$M_{A_2} = 1$
D	$M_{A_1} = 1$	$M_{A_2} \neq 1$
E	$M_{A_1} = M_1^+$	$M_{A_2} = M_2^+$

These points correspond, respectively, to the following particular cases: A, slow magnetoacoustic wave; B, switch-off shock; C, Alfvén shock; D, switch-on shock for the reversed flow; E, fast magnetoacoustic wave.

If we invert the direction of the flow, the state (2) becomes the upstream state and state (1) the downstream state; i.e., the parts of the curve in this line which correspond to entropy increase, and point B corresponds to the switch-on shock for this flow.

To construct the shock polar, we can introduce the variables

$$\xi = M_{A_1}^2 - 1, \qquad \eta = M_{A_2}^2 - 1$$

and write

$$q = \frac{(1 - M_1^{-2})(M_1^{+2} - 1)}{\gamma - 1}$$

$$r = (2 - \gamma)q$$

$$p = r - 2\frac{M_1^{-2} + M_1^{+2} - 2}{\gamma - 1}$$

The equation of the shock polar thus becomes

$$\xi = \eta\frac{[(\gamma + 1)/(\gamma - 1)]\,\eta^2 + p\eta - q}{\eta^2 + r\eta + q} \tag{3.1.5-5}$$

3.1.6. Detonations and Deflagrations

In MFD as in classical aerodynamics we are obliged to consider, alongside ordinary shock waves, those called combustion waves. The gas going through a combustion wave is subject to a chemical reaction (combustion) which has the effect of changing the equation of state $p = p(p, T)$ and hence of changing the specific internal energy $\epsilon_i(\tau, s)$. We will represent combustion phenomena by using two different expressions for the function $\epsilon_i(\tau, s)$: $\epsilon_i^{(1)}(\tau, s)$, the specific internal energy of the unburnt gas (ahead of the shock), and $\epsilon_i^{(2)}(\tau, s)$ the specific internal energy of the burnt gas (behind the shock).

We can distinguish two types of discontinuities:

$$\epsilon_i^{(1)}(\tau, s) - \epsilon_i^{(2)}(\tau, s) > 0, \qquad \text{exothermic discontinuity}$$

$$\epsilon_i^{(1)}(\tau, s) - \epsilon_i^{(2)}(\tau, s) < 0, \qquad \text{endothermic discontinuity}$$

The equations of combustion shocks can be deduced from the shock equations changing ϵ_i to $\epsilon_i^{(1)}$ in front of the shock and ϵ_i by $\epsilon_i^{(2)}$ behind the shock. It follows that the plane $(\mathbf{n}, \mathbf{H})$ is still invariant across the shock and that the component of $\mathbf{V}$ perpendicular to $(\mathbf{n}, \mathbf{H})$ is likewise invariant and hence could be taken equal to zero, as in Section 3.1.2. We then obtain the following equations (with $V = 0$):

$$\begin{gathered}
\rho_1 V_{n_1} = \rho_2 V_{n_2} = G \\
\rho_1 V_{n_1}^2 + p_1 + \frac{\mu h_1{}^2}{2} = \rho_2 V_{n_2}^2 + p_2 + \frac{\mu h_2{}^2}{2} = F_n \\
h_1(\tau_1 - \tau^*) = h_2(\tau_2 - \tau^*) = F_t \\
V_{n_1} h_1 - H_n v_1 = V_{n_2} h_2 - H_n v_2 = E \\
\epsilon_i^{(1)} + p_1\tau_1 + \frac{G^2\tau^2}{2} + \frac{h^2\tau^2}{2\tau} = \epsilon_i^{(2)} + p_2\tau_2 + \frac{G^2\tau^2}{2} + \frac{h^2\tau^2}{2\tau} = K
\end{gathered} \tag{3.1.6-1}$$

The quantities G, F_n, F_t, E, and K are constants. Expressing V_n, v, and h as functions of $\tau = 1/\rho$, we obtain the following two relations:

$$\begin{aligned}
&(\mathscr{F}) \qquad G^2\tau + p + \frac{\mu F_t{}^2}{2(\tau - \tau^*)^2} = F_n \\
&(\mathscr{C}_i) \qquad \epsilon_i^{(\alpha)}(p, \tau) + p\tau + \frac{G^2\tau^2}{2} + \frac{\mu F_t{}^2\tau^2}{2\tau(\tau - \tau^*)^2} = K
\end{aligned} \tag{3.1.6-2}$$

Hence, in the plane (τ, p), the states upstream of the shock are represented by the intersections (S_i) of the curves $(\mathscr{F})$ and $(\mathscr{C}_1)$, the states downstream of the shock are represented by the intersection (S_i') of the curves $(\mathscr{F})$ and $(\mathscr{C}_2)$.

In practical applications, we are led to take

$$\epsilon_{\text{i}}^{(1)}(p, \tau) = \frac{1}{\gamma_1 - 1} \frac{p}{\rho} + C_1, \qquad C_1 = \text{const}$$

$$\epsilon_{\text{i}}^{(2)}(p, \tau) = \frac{1}{\gamma_2 - 1} \frac{p}{\rho} + C_2, \qquad C_2 = \text{const}$$

There therefore exist for given values of the constants G, F_{n}, F_{t}, E, and K at most four points S_i and four points (S_i').

Since we have no information initially about the values of the specific entropy at the points (S_i) and (S_i'), 16 types of discontiniuties are possible:

$$S_i \to S_j', \qquad i, j = 1, 2, 3, 4$$

In the case of nonreacting media, the curves $(\mathscr{C}_1)$ and $(\mathscr{C}_2)$ are the same, and discontinuities $(S_i) \to (S_i')$ do not exist.

We assume that the points (S_i) and (S_j) are arranged in the order of increasing specific entropy:

$$s(S_i) < s(S_j) \qquad \text{and} \qquad s(S_i') < s(S_j') \qquad \text{for} \quad i < j$$

We call the following eight discontinuities intermediate shocks:

$$S_\alpha \to S_\beta' \qquad \text{or} \qquad S_\beta \to S_\alpha' \qquad (\alpha = 1, 2; \beta = 3, 4)$$

and we give the other eight discontinuities the following definitions:

$S_1 \to S_2'$, fast, strong detonation

$S_1 \to S_1'$, fast, weak detonation

$S_2 \to S_1'$, fast, strong deflagration

$S_2 \to S_2'$, fast, weak deflagration

$S_3 \to S_4'$, slow, strong detonation

$S_3 \to S_3'$, slow, weak detonation

$S_4 \to S_3'$, slow, strong deflagration

$S_4 \to S_4'$, slow, weak deflagration

The terminology adopted was chosen so that we recover the results concerning combustion waves of classical aerodynamics when the magnetic field is zero, and, on the other hand, those concerning the shocks of MFD (slow and fast shocks, which will be defined in Section 3.2.2) when the medium is not reactive.

The inequalities (3.1.4-4) are still satisfied such that the normal velocities in front of and behind the shock satisfy the following fundamental inequalities:

Fast, strong detonation $U_+^{(1)} \leqslant |V_n^{(1)} - U|,$

$U_A^{(2')} \leqslant |V_n^{(2')} - U| \leqslant U_+^{(2')}$

Fast, weak detonation $U_+^{(1)} \leqslant |V_n^{(1)} - U|,$

$U_+^{(1')} \leqslant |V_n^{(1')} - U|$

Fast, strong deflagration $U_A^{(2)} \leqslant |V_n^{(2)} - U| \leqslant U_+^{(2)},$

$U_+^{(1')} \leqslant |V_n^{(1')} - U|$

Fast, weak deflagration $U_A^{(2)} \leqslant |V_n^{(2)} - U| \leqslant U_+^{(2)},$

$U_A^{(2')} \leqslant |V_n^{(2')} - U| \leqslant U_+^{(2')}$

Slow, strong detonation $U_-^{(3)} \leqslant |V_n^{(3)} - U| \leqslant U_A^{(3)},$

$|V_n^{(4')} - U| \leqslant U_-^{(4')}$

Slow, weak detonation $U_-^{(3)} \leqslant |V_n^{(3)} - U| \leqslant U_A^{(3)},$

$U_-^{(3')} \leqslant |V_n^{(3')} - U| \leqslant U_A^{(3')}$

Slow, strong deflagration $|V_n^{(4)} - U| \leqslant U_-^{(4)},$

$U_-^{(3')} \leqslant |V_n^{(3')} - U| \leqslant U_A^{(3')}$

Slow, weak deflagration $|V_n^{(4)} - U| \leqslant U_-^{(4)},$

$|V_n^{(4')} - U| \leqslant U_-^{(4')}$

3.2. Stability of Shock Waves

3.2.1. The Choice of a Criterion of Stability.

The definition of stability in problems of mechanics always contains an element of arbitrariness; the conclusions differ according to the criterion chosen. The problem which we will study is a particular problem of shock-wave stability.

We consider a plane, steady shock wave; we take the $y0z$ plane as the plane of the shock and the $0x$ axis as a line perpendicular to this plane directed so that the component of velocity along the axis is positive. The shock wave divides the fluid into two regions: region (1) ($x < 0$) and region (2) ($x > 0$). We assume that the flow is uniform on both sides of the wave; the quantities defining the state of the fluid have constant values in region (1) and different constant values in region (2). These quantities are the components of the magnetic field, of the velocity, the specific density, and specific entropy; we will regard these quantities as the components of a vector $\mathbf{z}$ in an n-dimensional Euclidean space. The components of $\mathbf{z}$ are discontinuous across the shock, but the discontinuities satisfy n conditions. The vector $\mathbf{z}$ has a constant value $\mathbf{Z}_1$ in region (1) and a constant value $\mathbf{Z}_2$ in region (2). At a given time, taken as the initial instant, we assume that the flow is slightly disturbed, the disturbances being functions only of the distance from the shock.

After the initial instant, the fluid will have a motion defined by $\mathbf{z}(t, x)$; this is a solution of the MFD equations, which can be written in the form

$$\frac{\partial \mathbf{z}}{\partial t} + A(\mathbf{z}) \frac{\partial \mathbf{z}}{\partial x} = 0 \qquad (3.2.1\text{-}1)$$

where the matrix $A(\mathbf{z})$ denotes a known function of the vector $\mathbf{z}$. When the initial motion is stable, the perturbations remain small in both regions. We write

$$\delta\mathbf{z}^{(1)}(t, x) = \mathbf{z}^{(1)}(t, x) - \mathbf{Z}_1$$

$$\delta\mathbf{z}^{(2)}(t, x) = \mathbf{z}^{(2)}(t, x) - \mathbf{Z}_2$$

The linearization of Eq. (3.2.1-1) in both regions yields the following linear homogeneous equation:

$$\frac{\partial(\delta\mathbf{z}^{(i)})}{\partial t} + A(\mathbf{Z}_i) \frac{\partial(\delta\mathbf{z}^{(i)})}{\partial x} = 0 \qquad (i = 1, 2) \qquad (3.2.1\text{-}2)$$

In our problem, the boundary conditions related to Eq. (3.2.1-2) are the following:

1. At the initial instant, $\mathbf{z}(t, x) = \mathbf{z}(0, x)$, a given function.
2. At infinity, the vector $\mathbf{z}(t, x)$ takes the given constant values $\mathbf{z}(t, -\infty) = \mathbf{Z}_1$ and $\mathbf{z}(t, +\infty) = \mathbf{Z}_2$.
3. On the shock wave, the values $\mathbf{z}^{(1)}(t, x)$ and $\mathbf{z}^{(2)}(t, x)$ are related by the shock equations.

Naturally, we assume that $\mathbf{z}(0, -\infty) = \mathbf{Z}_1$ and $\mathbf{z}(0, +\infty) = \mathbf{Z}_2$†. The search for a solution of Eq. (3.2.1-1) satisfying the preceding conditions constitutes a boundary value problem, which will be called problem $\mathscr{P}$. We suppose that problem $\mathscr{P}$ has one and only one solution. When the flow represented by this solution is stable, the linearization is valid and problem $\mathscr{P}$ applied to the linearized equation

$$\frac{\partial(\delta\mathbf{z})}{\partial t} + A(\mathbf{Z}) \frac{\partial(\delta\mathbf{z})}{\partial x} = 0 \tag{3.2.1-3}$$

likewise has a unique solution. This result forms the basis of the following choice of stability criteria: The flow with a shock wave, which we have set out, is said to be stable if the $\mathscr{P}$ problem connected with the linearized equation (3.2.1-3) has a unique solution. It is said to be unstable otherwise.

Since Eq. (3.2.1-3) is linearized, it can be integrated by means of the Laplace transform. For this, we introduce the function

$$\mathscr{T}(\tau, x) = \int_0^\infty e^{-\tau t}\, \delta\mathbf{z}(t, x)\, dt, \qquad \tau > 0$$

Integration by parts gives

$$\tau\mathscr{T}(\tau, x) = \int_0^\infty e^{-\tau t} \frac{\partial[\delta\mathbf{z}(t, x)]}{\partial t}\, dt - \delta\mathbf{z}(t, x)\, e^{-\tau t} \Big|_0^\infty$$

We assume that $\delta\mathbf{z}(t, x)$ is bounded when t is infinite, so that the last term of the r.h.s. reduces to $\delta\mathbf{z}(0, x)$. Equation (3.2.1-3) then reduces to

$$A(\mathbf{Z}) \frac{\partial\mathscr{T}}{\partial x} + \tau\mathscr{T} = \delta\mathbf{z}(0, x) \tag{3.2.1-4}$$

The homogeneous equation associated with Eq. (3.2.1-4) has a solution of the form $\mathscr{T}(\tau, x) = \boldsymbol{\alpha}\, e^{\lambda x}$, provided that $A\lambda\boldsymbol{\alpha} + \tau\boldsymbol{\alpha} = 0$, if the ratio $-(\tau/\lambda)$ is equal to one of the eigenvalues g_i of the matrix $A(\mathbf{Z})$; if the vector $\boldsymbol{\alpha}$ is equal to one of the corresponding eigenvectors $\mathbf{r}_i(\mathbf{Z})$, then the matrix A of rank n possesses n eigenvalues (real eigenvalues in the case of MFD). We put

$$\mathscr{T}_i(\tau, x) = \mathbf{r}_i[\exp(-\tau x/g_i)]\mathscr{T}, \qquad \tau > 0 \tag{3.2.1-5}$$

† We assume, in fact, that $\delta\mathbf{z}(0, x) = 0$, for $|x|$ large enough.

It is possible to look for a general solution of Eq. (3.2.1-4) in the form

$$\mathscr{T}(\tau, x) = \sum_{i=1}^{n} h_i(x)\, \mathscr{T}_i(\tau, x) \qquad (3.2.1\text{-}6)$$

Substitution in Eq. (3.2.1-4) gives

$$\sum_{i=1}^{n} \mathbf{r}_i \left[\exp\left(-\frac{\tau x}{g_i}\right)\right] \frac{dh_i}{dx} = A^{-1}\, \delta\mathbf{z}(0, x)$$

To solve this last equation, we denote by $\mathbf{H}$ the vector with components $[\exp(-\tau x/g_i)]\, dh_i/dx$, and by R the matrix, the elements of which are the components of the eigenvectors $\mathbf{r}_i$, we then have

$$R\,\mathbf{H} = A^{-1}\, \delta\mathbf{z}(0, x)$$

$$\mathbf{H} = R^{-1}A^{-1}\, \delta\mathbf{z}(0, x)$$

$$\frac{dh_i}{dx} = \left[\exp\left(\frac{\tau x}{g_i}\right)\right] \{R^{-1}A^{-1}\, \delta\mathbf{z}(0, x)\}_i$$

The last expression denotes the component of rank i of the vector $R^{-1}A^{-1}\, \delta\mathbf{z}(0, x)$. We then obtain

$$h_i(x) = h_i(0) + \int_0^x [\exp(\tau\xi/g_i)]\{R^{-1}A^{-1}\, \delta\mathbf{z}(0, \xi)\}_i\, d\xi \qquad (3.2.1\text{-}7)$$

When the functions $h_i(x)$ are determined in this way, the solution given by formula (3.2.1-6) satisfies the initial conditions. It still contains $2n$ arbitrary constants, $h_i^{(1)}(0)$ and $h_i^{(2)}(0)$, $i = 1, 2, \ldots, n$.

The conditions at infinity require that the perturbations tend to zero for all t. The region at infinity corresponds to $x = -\infty$ in region (1) and to $x = +\infty$ in region (2). The behavior of the vectors $\mathscr{T}_i(\tau, x)$ at infinity depends on the sign of the eigenvalue g_i; at infinity, $\mathscr{T}_i(t, x)$ is not bounded if $g_i^{(1)} > 0$ and $g_i^{(2)} < 0$; and $\mathscr{T}_i(t, x)$ is zero if $g_i^{(1)} < 0$ and $g_i^{(2)} > 0$.

Conditions at infinity require that the coefficients $h_i^{(1)}(-\infty)$ and $h_i^{(2)}(+\infty)$ which correspond to the first case are zero; the corresponding constants of integration $h_i(0)$ are hence determined. The other constants of integration are still arbitrary; their number m is equal to the number of eigenvalues of A such that either $g_i^{(1)} < 0$ or $g_i^{(2)} > 0$.

The solution of the $\mathscr{P}$ problem will be unique when the shock-wave conditions permit us to compute all the m constants uniquely. The shock-wave conditions relate the components of the vectors $\mathbf{z}^{(1)}$ and $\mathbf{z}^{(2)}$; the number of these relations is equal to the number n of the components of the vector $\mathbf{z}$; since the shock velocity is included in these relations, we obtain after its elimination $(n - 1)$ relations between the components of the vectors $\mathbf{z}^{(1)}$ and $\mathbf{z}^{(2)}$ for $x = 0$. Hence, for $m < n - 1$, the $\mathscr{P}$ problem is impossible, while for $m > n - 1$, the $\mathscr{P}$ problem is indeterminate.

For the flow under consideration (flow with shock waves) to be stable, we must have $m = n - 1$. We also require that the condition at the shock determine uniquely those m constants of integration which are not fixed by conditions at infinity.

To apply the method just shown to concrete problems, it is convenient to give a physical interpretation of the number m. The computations of the previous chapter show that the eigenvalues g_i of the matrix A represent displacement velocities of linearized waves (surfaces of discontinuities for the derivatives of the components of $\mathbf{z}$). Each wave progressing in the positive direction corresponds to positive eigenvalues, and conversely (Fig. 3.2.1.a).

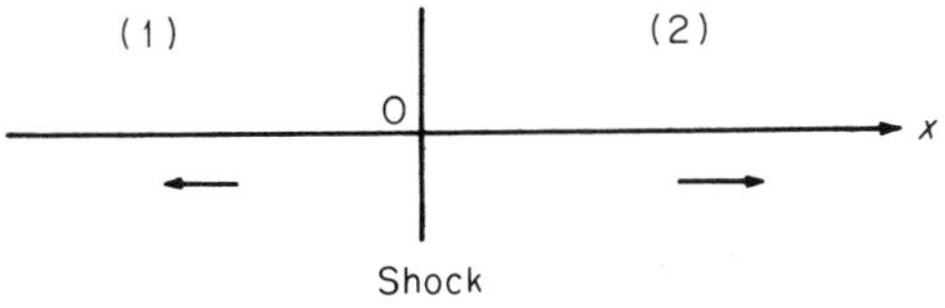

FIG. 3.2.1.a. Correspondence between wave direction and eigenvalue sign. (1) $g_i^{(1)} < 0$; (2) $g_i^{(2)} > 0$.

The number m of the eigenvalues satisfying either $g_i^{(1)} < 0$ or $g_i^{(2)} > 0$ is hence equal to the number of diverging waves, waves that move away from the plane $x = 0$. The waves moving toward the plane $x = 0$ are called converging waves. Hence, the study of the $\mathscr{P}$ problem reduces to the study of the number of diverging waves.

3.2.2. Applications to MFD

The linearized waves of MFD can be propagated with one of the four speeds U_A, U_-, U_+, and speed zero. In the case of one-dimensional flow, all the waves are plane, perpendicular to $0x$, and propagate along $0x$; we then have

$$U_A{}^2 = \frac{\mu H_x{}^2}{\rho}$$

$$2U_\pm{}^2 = \frac{\partial p}{\partial \rho} + \frac{\mu H^2}{\rho} \pm \left\{\left(\frac{\partial p}{\partial \rho} + \frac{\mu H^2}{\rho}\right)^2 - 4\frac{\partial p}{\partial \rho}\frac{\mu H_x{}^2}{\rho}\right\}^{1/2}$$

To the four propagation speeds, there correspond seven displacement speeds: V_x, entropy waves; $V_x + U_A$ and $V_x - U_A$ Alfvén waves; $V_x + U_-$ and $V_x - U_-$, slow magnetoacoustic waves; and $V_x + U_+$ and $V_x - U_+$, fast magnetoacoustic waves.

The waves which move with one of the velocities V_x, $V_x + U_A$, $V_x + U_-$, and $V_x + U_+$ are always convergent in region (1) and divergent in region (2). The waves moving with one of the speeds $V_x - U_A$, $V_x - U_-$, and $V_x - U_+$ can be either convergent or divergent. There are six of the latter, three for each region; among these six, we call m' the number of the divergent ones; we have $m = m' + 4$.

The vector $\mathbf{z}$ has seven components. In fact, the unknowns of the problem are *a priori*: the vectors $\mathbf{H}$ and $\mathbf{V}$ and the scalars ρ and s, which together make eight unknowns. But the equation $\partial\mathbf{H}/\partial t - \nabla \times (\mathbf{V} \times \mathbf{H}) = 0$, noting that we seek only perturbations which are functions of t and x, allows us to write

$$\partial H_x/\partial t = 0, \qquad \text{where} \quad H_x(t, x) = H_x(0, x) = \text{const}$$

so that the component of the magnetic field normal to the shock is not an unknown, and we have $n = 7$.

The necessary conditions for stability, $m = n - 1$, gives us our case $m' = 2$. In other words, two of the six waves which propagate in either region with one of the speeds $V_x - U_A$, $V_x - U_-$, and $V_x - U_+$ must be divergent, the other four being convergent. For a wave to be divergent, it must travel with negative velocity in region (1) and positive velocity in region (2). Since the propagation speeds satisfy the conditions $U_- \leqslant U_A \leqslant U_+$, we have 16 possibilities (not all realizable), shown in Fig. 3.2.2.a.

If the velocity $V_x^{(1)}$ ahead of the shock is smaller than $U_-^{(1)}$, i.e., in one of the four cases on the left, there are three divergent waves toward the left. If $V_x^{(1)}$ satisfies the inequality $U_-^{(1)} < V_x^{(1)} < U_A^{(1)}$, i.e., in the column immediately to the right of the preceding, then we

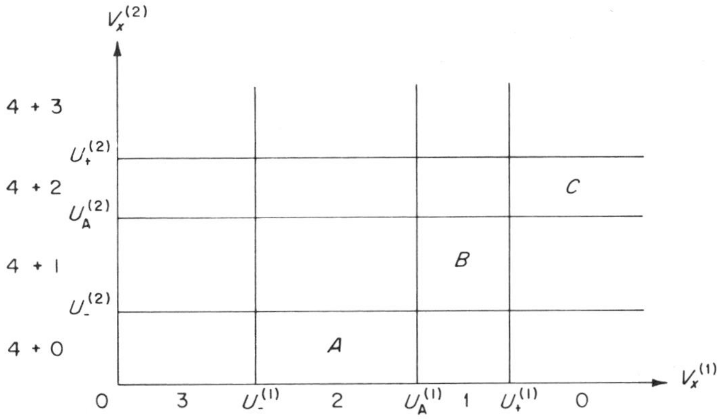

FIG. 3.2.2.a. Number of divergent waves.

have two divergent waves running to the left; the number of divergent waves running to the left is shown below each column. Similarly, the number of divergent, right-running waves is shown on the left of each line. The total number of divergent waves obtained when $V_x^{(1)}$ and $V_x^{(2)}$ satisfy one of the 16 inequalities is obtained by adding the number of left-running divergent waves (written below) and right-running divergent waves (written on the left); the total is equal to six only in the cases A, B, and C. Hence, the flow will be stable in one of the three following cases:

$$A\colon \quad U_-^{(1)} < V_x^{(1)} < U_A^{(1)} \quad \text{and} \quad V_x^{(2)} < U_-^{(2)}.$$

$$B\colon \quad U_A^{(1)} < V_x^{(1)} < U_+^{(1)} \quad \text{and} \quad U_-^{(2)} < V_x^{(2)} < U_A^{(2)}.$$

$$C\colon \quad U_+^{(1)} < V_x^{(1)} \quad \text{and} \quad U_A^{(2)} < V_x^{(2)} < U_+^{(2)}.$$

In fact, when the magnetic field, the pressure, and the density ahead of the shock are given, $U_-^{(1)}$, $U_A^{(1)}$, and $U_+^{(1)}$ are known. Once we know $V_x^{(1)}$, then $V_x^{(2)}$, $U_-^{(2)}$, $U_A^{(2)}$, and $U_+^{(2)}$ are determined by the shock equations, so that the 16 boxes of Fig. 3.2.2.a do not all correspond to the cases possible. But the cases A, B, and C are realizable. Moreover, the conditions for stability just derived contain only velocities, namely, the gas velocities and velocities of propagation of linearized waves. These conditions are not sufficient.

The state of motion in front of and behind the shock satisfy the shock equations for both the disturbed and undisturbed flow. The equations of conservation of mass give, for example,

$$\rho_1(V_{n_1} - U) = \rho_2(V_{n_2} - U)$$

$$(\rho_1 + \delta\rho_1)(V_{n_1} + \delta V_{n_1} - U - \delta U) = (\rho_2 + \delta\rho_2)(V_{n_2} + \delta V_{n_2} - U - \delta U)$$

Subtracting and keeping only the linear terms, we get

$$[\delta\rho(V_n - U) + \rho(\delta V_n - \delta U)] = 0$$

where $[Q]$ means $Q_2 - Q_1$. Applying similar arguments to the other shock equations, we get a set of relations satisfied by the perturbed flow at the shock. These relations are as follows:

$$[(V_x - U)\,\delta H_y - H_x\,\delta V_y + (\delta V_x - \delta U)H_y] = 0 \quad (3.2.2\text{-1a})$$

$$[(V_x - U)\,\delta H_z - H_x\,\delta V_z + \cancel{(\delta V_x - \delta U)H_z}] = 0 \quad (3.2.2\text{-1b})$$

$$[(V_x - U)\,\delta\rho + (\delta V_x - \delta U)\rho] = 0 \quad (3.2.2\text{-1c})$$

$$[\rho(V_x - U)\,\delta V_y - \mu H_x\,\delta H_y] = 0 \quad (3.2.2\text{-1d})$$

$$[\rho(V_x - U)\,\delta V_z - \mu H_x\,\delta H_z] = 0 \quad (3.2.2\text{-1e})$$

$$[\rho(V_x - U)\,\delta V_x + \delta p + \mu(H_y\,\delta H_y + \cancel{H_z\,\delta H_z})] = 0 \quad (3.2.2\text{-1f})$$

$$[\rho(V_x - U)\,\delta\epsilon_i + p\,\delta V_x + (\tfrac{1}{2}\mu H^2 - \mu H_x^2)\,\delta U] = 0 \quad (3.2.2\text{-1g})$$

In the above, V_x is the only nonzero component of the velocity, and U is the velocity of propagation of the shock in the undisturbed flow. We choose one coordinate system so that $U = 0$; δU will then mean the velocity of propagation of the shock in the disturbed flow.

Equations (3.2.2-1a-g) determine the quantities δH_y, δH_z, δV_x, δV_y, δV_z, δp, and $\delta\rho$ on one side of the shock once we know these values on the other side of the shock together with the shock velocity $\delta U - \delta V_x$; ϵ_i is a known function of p and ρ. Since the plane $(\mathbf{n}, \mathbf{H})$ is invariant across the shock, we can assume $H_z = 0$ (ahead of as well as behind the shock). Hence, the system (3.2.2-1a-g) resolves into two independent systems: 1. Equations (3.2.2-1b, e) containing only δV_z and δH_z. 2. Equations (3.2.2-1a, c, d, f, g), containing only δH_y, δV_x, δV_z, δp, $\delta\rho$, and δU.

The discontinuities of δH_z and δV_z propagate with the Alfvén speed and are therefore carried by Alfvén waves. In fact, if we make the computations again, with $\delta H_y = \delta V_x = \delta V_y = \delta p = \delta S = 0$, we get $\rho(V_x - U)^2 = \mu H_x^2$.[1]

In a similar fashion, the discontinuities of δH_y, δV_x, δV_y, δp, and $\delta \rho$ propagate with zero velocity or with magnetoacoustic speed and are carried by an entropy wave or by a magnetoacoustic wave. On the other hand, Alfvén waves carry discontinuities in δH_z and δV_z only, the only components of the eigenvector $\mathbf{r}$ when $H_z = 0$. Entropy or magnetoacoustic waves only carry discontinuities in δH_y, δV_x, δV_y, δp, and $\delta \rho$. On a magnetoacoustic or entropy wave, we have five relations; these relations contain the perturbed shock speed δU; there then exist four independent relations between the values of δH_y, δV_x, δV_y, δp, and $\delta \rho$ on each side of the shock. To determine the corresponding unknown constants $h_i(0)$, they must be four in number, that is, there are four diverging magnetoacoustic or entropy waves. On an Alfvén wave, we have two relations which do not contain δU; there are therefore two independent relations to determine the values of δV_z and δH_z on each side of the shock wave. To determine the constants $h_i(0)$, they must be two in number, that is, there are two diverging Alfvén waves.

Hence, the stability conditions stated (namely, that the number of divergent waves is equal to six) enable us to determine the arbitrary constants $h_i(0)$ and will therefore be sufficient conditions for stability if the number of divergent waves is equal to four, in the case of entropy or magnetoacoustic waves, and two for the Alfvén waves. Figure 3.2.2.b gives a summary of the results; in each box is shown the number of divergent magnetoacoustic and entropy waves (first digit) and the number of divergent Alfvén waves (second digit). Only in

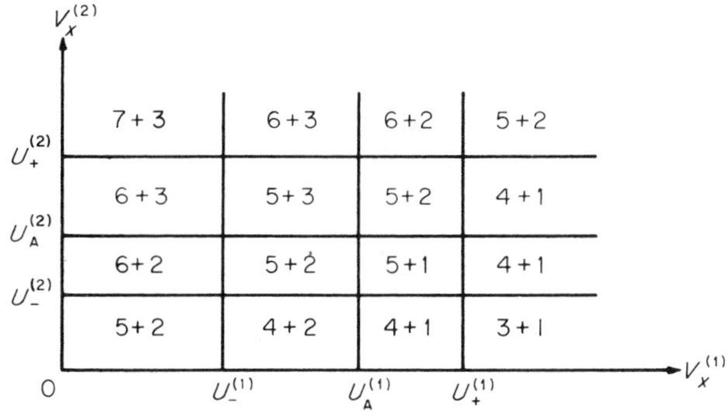

FIG. 3.2.2.b. Number of different types of divergent waves.

[1] Because, for $H_y = 0$, Eq. (2.2.3-1) splits into two independent systems.

the regions A and C, where these numbers are 4 and 2, can we expect stability. Region B is an unstable region even if the total number is 6, for, in this region, we have five magnetoacoustic or entropy waves and one Alfvén wave.

3.2.3. Consequences of the Stability Conditions

The only shock waves observed in practice are those corresponding to stable flow, i.e., those satisfying the inequalities A or C of Section 3.2.2 using a shock coordinate system. If we now use another frame of reference in which the shock has a velocity D, inequalities A and C will have to be rewritten

$$A:\quad U_-^{(1)} < |V_x^{(1)} - D| < U_A^{(1)} \quad \text{and} \quad |V_x^{(2)} - D| < U_-^{(2)}$$

$$C:\quad U_+^{(1)} < |V_x^{(1)} - D| \quad \text{and} \quad U_A^{(2)} < |V_x^{(2)} - D| < U_+^{(2)}$$

The index (1) denotes the region upstream of the shock and (2) the region downstream. The two terms $V_x^{(1)} - D$ and $V_x^{(2)} - D$ have the same sign because

$$\rho_1(V_x^{(1)} - D) = \rho_2(V_x^{(2)} - D)$$

and

$$|V_x^{(2)} - D| < |V_x^{(1)} - D|$$

since $\rho_2 > \rho_1$.

The shocks of the first type are slow (shocks $3 \rightarrow 4$) as defined in Section 3.1; the shocks of the second type are fast (shocks $1 \rightarrow 2$). The speed of propagation $|V_x - D|$ of a slow shock is less than the Alfvén speed both ahead of and behind the shock, while that of a fast shock is greater than the Alfvén speed.

We recall that Alfvén shocks are defined as those for which $V_x^{(1)} = U_{\mathrm{A}}^{(1)} = U_{\mathrm{A}}^{(2)} = V_x^{(2)}$.

The shock-stability conditions lead to the following interesting results.

Theorem 1. If two slow (or fast) shocks follow each other, the shock behind always catches up with the one ahead (Fig. 3.2.2.c).

We will prove this in the case of two slow shocks. The two parallel slow shocks divide the whole space of interest into three regions: region α, in which the fluid is at rest and has not passed through any shock; this region extends from the first shock which moves with a

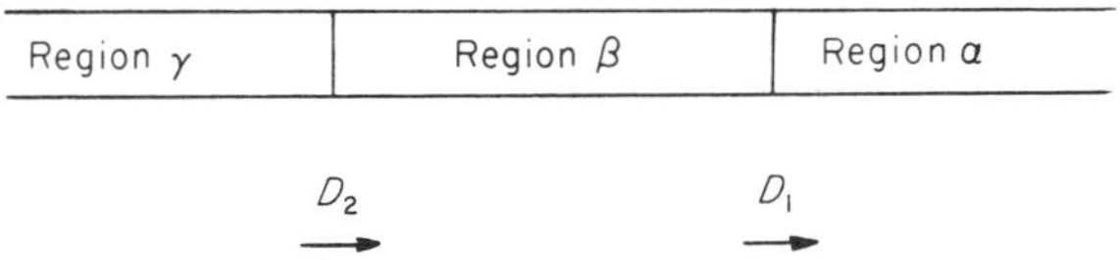

FIG. 3.2.2.c. First shock, speed D_1. Second shock, speed D_2.

positive speed D_1 to $+\infty$; region β, in which the fluid has crossed the first shock but not the second; this region lies between the two shocks, the second traveling with speed $D_2 > V_x^{(\beta)}$; and region γ, in which the fluid has crossed both shocks. The inequalities A are satisfied at each of the two shocks. They can be written for the first shock as

$$U_-^{(\alpha)} < D_1 < U_A^{(\alpha)} \qquad \text{and} \qquad D_1 - V_x^{(\beta)} < U_-^{(\beta)}$$

and for the second shock as

$$U_-^{(\beta)} < D_2 - V_x^{(\beta)} < U_A^{(\beta)} \qquad \text{and} \qquad D_2 - V_x^{(\gamma)} < U_-^{(\gamma)}$$

In particular, we deduce that

$$D_1 - V_x^{(\beta)} < U^{(\beta)} < D_2 - V_x^{(\beta)}$$

Hence, $D_1 < D_2$, which proves that the second shock catches up with the first.

Theorem 2. An Alfvén shock always catches up with a slow shock. A fast shock always overtakes an Alfvén shock. A fast shock always catches up with a slow shock.

We again make use of Fig. 3.2.2.c, assuming that $V_x^{(\alpha)} = 0$, $D_1 > 0$, and also assume that the first shock is slow while the second is an Alfvén shock; we have, for the first shock,

$$U_-^{(\alpha)} < D_1 < U_A^{(\alpha)} \qquad \text{and} \qquad D_1 - V_x^{(\beta)} < U_-^{(\beta)}$$

and for the second shock,

$$D_2 - V_x^{(\beta)} = U_A^{(\beta)}$$

Hence,

$$D_1 - V_x^{(\beta)} < U_-^{(\beta)} < U_A^{(\beta)} = D_2 - V_x^{(\beta)}$$

The case of several Alfvén shocks succeeding each other need not be considered since they propagate at the same speed and their total effect is the same as that of single Alfvén shock. With several Alfvén shocks corresponding to one Alfvén shock, we can write, using the two previous theorems:

Theorem 3. At most three shocks can propagate together without merging; the first will be a fast, the second an Alfvén, and the third a slow shock.

Theorem 4. A fast shock always catches up with any linearized wave. An Alfvén shock always overtakes a slow magnetoacoustic wave. A slow shock always overtakes a slow magnetoacoustic wave.

Referring again to Fig. 3.2.2.c, we assume, for example, that the first shock is replaced by a linearized wave (its velocity is $U_+^{(\alpha)}$ at the most) and that the second shock is fast. For the linearized wave, we have

$$D_1 - V_x^{(\alpha)} = D_1 - V_x^{(\beta)} < U_+^{(\alpha)} = U_+^{(\beta)}$$

For the shock,

$$U_+^{(\beta)} < D_2 - V_x^{(\beta)} \qquad \text{and} \qquad U_A^{(\gamma)} < D_2 - V_x^{(\gamma)} < U_+^{(\gamma)}$$

Hence,

$$D_1 - V_x^{(\beta)} = U_+^{(\beta)} < D_2 - V_x^{(\beta)}$$

Theorem 5. A fast magnetoacoustic wave catches up with any shock. A slow magnetoacoustic wave always overtakes a slow shock.

Theorem 6. The transverse component of the magnetic field (the direction of which is conserved) increases in a fast shock and decreases in a slow shock.

This result is a consequence of Eq. (3.1.1-4), which can also be written

$$\mathbf{h}_2 = \mathbf{h}_1 \frac{\tau_1 - \tau^*}{\tau_2 - \tau^*} \qquad \text{with} \quad \tau^* = \frac{\mu H_x{}^2}{\rho^2(V_x - D)^2}$$

with the property that shocks are compressive, giving $\tau_2 < \tau_1$. The square of the Alfvén speed will have the value

$$U_A{}^2 = \mu H_x{}^2/\rho = (\tau^*/\tau)(V_x - D)^2$$

In a fast shock, we have, in particular, $U_A^{(2)} < | V_x^{(2)} - D |$; hence, $\tau^* < \tau_2$. We can deduce that $\tau_1 - \tau^* > \tau_2 - \tau^* > 0$; hence,

$h_2 > h_1$. In a slow shock, we have, in particular, $| V_x^{(1)} - D | < U_A^{(1)}$; hence, $\tau^* > \tau_1$. We deduce that $\tau^* - \tau_2 > \tau^* - \tau_1 > 0$; hence, $h_2 < h_1$.

Recall that, in an Alfvén shock, the tangential component is conserved in magnitude but not in direction. Since the normal magnetic field is conserved, we can write Theorem 6 as:

Theorem 7. The intensity of the magnetic field increases across a fast shock and decreases across a slow shock.

3.2.4. Stability of Detonations and Deflagrations

The theory of stability presented in Section 3.2.1 can be applied with equal force to combustion waves. The linearized shock conditions form a system of seven equations which can be split into two independent systems; a system of five equations in which the perturbed shock velocity appears, and a second of two equations in which this velocity does not appear. The number of divergent waves corresponding to each of these systems is shown on Fig. 3.2.4.a.

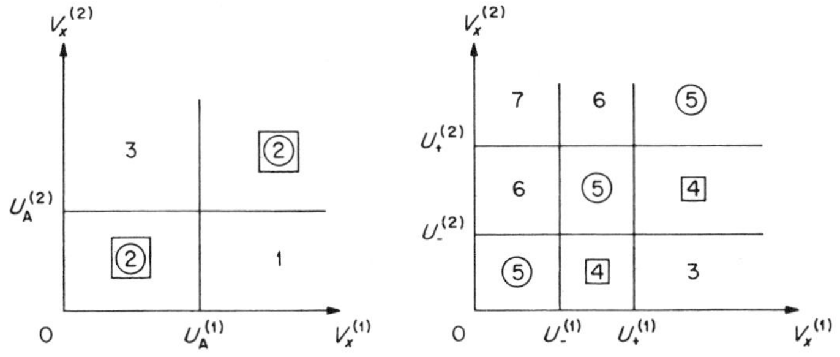

FIG. 3.2.4.a. Left: number of divergent Alfvén waves. Right: number of divergent magnetoacoustic and entropy waves.

The corresponding stability conditions for the system of two equations, since this does not contain the shock speed, is that the number of divergent waves equals two. The system of five equations contains the shock velocity, but, as we will see in Chapter 4, this velocity is unknown for some combustion waves; it must therefore be calculated from the shock equations. However, for other combustion waves, weak detonations and deflagrations, the velocity is known from physical and chemical properties of the gas. Consequently we must distinguish between two cases:

1. When the shock velocity is unknown, the shock equations yield four equations independent of it, and the condition for stability is that the number of divergent waves equals four. The cases in which waves of this type are stable are marked by a square in Figs. 3.2.4.a and 3.2.4.b.

2. When the shock velocity is known, the shock equations yield five relations and the condition for stability is that the number of divergent waves be five. The cases in which waves of this type are stable are indicated by circles in Figs. 3.2.4.a and 3.2.4.b.

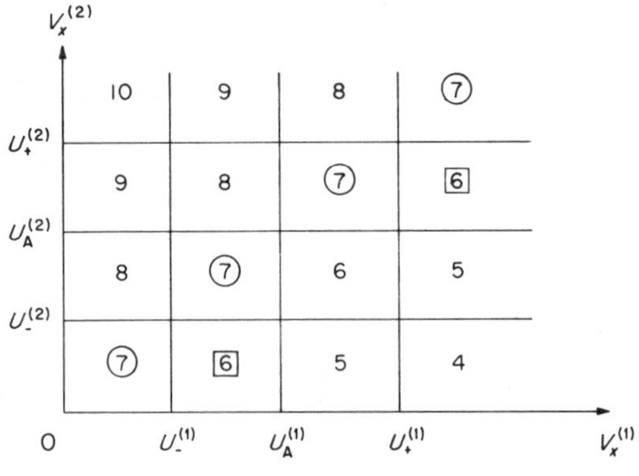

FIG. 3.2.4.b. Total number of divergent waves.

We thus find that there are six cases in which combustion waves can be stable. It is found that all six of these can be realized. In fact, the four cases shown by circles in Fig. 3.2.4.b correspond to weak detonations and deflagrations; we will show in Chapter 4 that for these the velocity of propagation depends on the physical and chemical properties of the gas. The cases shown by squares correspond to strong detonations. Strong deflagrations and eight other shocks which have not been assigned special names are unstable.

3.3. The Wedge and Piston Problems

3.3.1. THE WEDGE PROBLEM

In this section, we consider the problem of steady flow past a wedge with an attached shock wave at the leading edge. The flow is assumed to be inviscid, have zero coefficient of heat conduction, and negligible electrical resistance. To avoid complicating the problem too much, we will not consider the most general case, but will assume that, in the flow far upstream, that is, ahead of all shocks, the velocity vector and the magnetic field are perpendicular to the leading edge of the wedge. The flow is then two dimensional (Fig 3.3.1.a). We

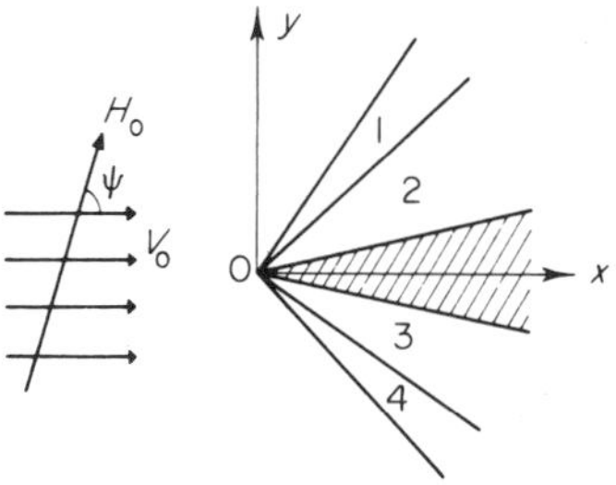

Fig. 3.3.1.a. Steady flow past a wedge, two-dimensional case.

assume that a certain number n of plane shocks, attached to the leading edge, are formed. These shocks and the wedge determine n regions, apart from the undisturbed regions far upstream where quantities are identified by subscript 0. In each of these regions, the flow and the magnetic field are uniform.

In each region, the flow is defined by six unknowns (two components of the magnetic field, two velocity components, pressure, and density). The boundary conditions to be satisfied are:

1. The shock equations (six on each wave).
2. The condition that the normal component of velocity is zero on the wedge surface.
3. Two conditions relative to the magnetic field in the regions adjacent to the wedge surfaces.

When the fluid and the wedge have the same magnetic permeability, the last two conditions state that the magnetic fields are equal in the regions next to the wedge. In fact, across the wedge boundary, the normal component of magnetic induction B_n is continuous (from Faraday's law), and so is the tangential component of magnetic field h [this follows from the second equation of (3.1.1-3) in which $V_n - U = 0$). The fact that the magnetic permeabilities are equal then implies that the magnetic field $\mathbf{H}$ is continuous. The wedge is assumed to have zero electric conductivity, so that the electric current is zero; we then have within the wedge

$$\nabla \times \mathbf{H} = 0 \qquad \text{and} \qquad \nabla \cdot \mathbf{H} = 0$$

From Maxwell's equation $\mathbf{J} = \nabla \times \mathbf{H}$ and from the initial condition $\nabla \cdot \mathbf{H} = 0$. In Fig. 3.3.1.a, which is orthogonal to the leading edge, we take the axes $0x$ along the bisector of the wedge angle and the axis $0y$ normal to this. The above vector equations thus become

$$\frac{\partial H_x}{\partial y} - \frac{\partial H_y}{\partial x} = 0 \qquad \text{and} \qquad \frac{\partial H_x}{\partial x} + \frac{\partial H_y}{\partial y} = 0$$

Using the Cauchy–Riemann relations, we can thus write $H_x - iH_y = f(x + iy)$, where f is an arbitrary function of the complex variable $x + iy$.

The magnetic field is therefore constant in the whole wedge, since it is constant along a straight line (the sides of the wedge); it therefore has the same value in both fluid regions adjacent to the wedge.

There are then $6n + 4$ boundary conditions; there are seven unknowns for each of the n flow regions—namely, six quantities defining the flow field, and the angle of the shock forming the upstream limit of the region considered. If we suppose that the problem has a solution (without actually finding this), we determine the value of the number n from the condition that the number of unknowns is equal to the number of boundary conditions: $7n = 6n + 4$, so that $n = 4$. Four shocks arise, in general, two on each side of the wedge.

When the wedge angle is very small, the shocks reduce to magneto-acoustic waves. In each of the four regions 1, 2, 3, 4, the velocity of the fluid is equal to the free-stream velocity $\mathbf{V}_0$, and on each wave the normal velocity component must equal one of the magnetoacoustic speeds. We can determine the waves graphically in the following fashion:

Starting from the wedge apex 0 (see Fig. 3.3.1.b), we draw the vector $\mathbf{0A}$ equal to $\mathbf{V}_0$. With A as center, we construct the Friedrichs diagram corresponding to the conditions of the problem; the support of the vector with origin A and equal to the magnetic field $\mathbf{H}_0$ is the axis of symmetry. The tangents to the Friedrichs diagram, drawn from 0, are the magnetoacoustic waves we are seeking, because, from the construction of the Friedrichs diagram, the projection of the

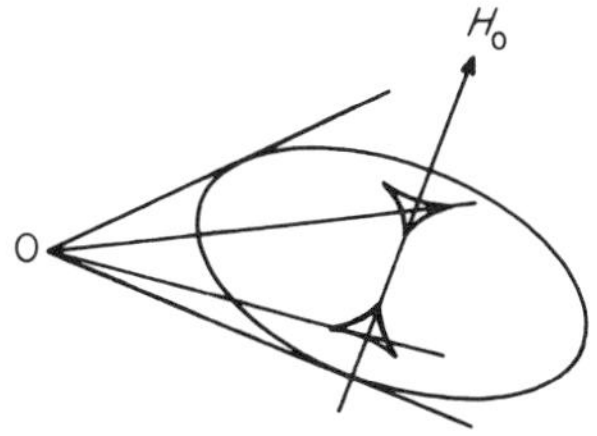

FIG. 3.3.1.b. Graphical determination of magnetoacoustic waves.

velocity $\mathbf{V}_0 = \mathbf{0A}$ on the normal to the wave is in fact equal to one of the magnetoacoustic speeds. When 0 is outside the corner curve of the Friedrichs diagram, the first two shocks are fast shocks and the last two are slow.

When the permeabilities of the fluid and of the wedge are different, the preceding results are modified somewhat; the magnetic fields in the regions (2) and (3) are no longer equal, but are connected by two relations, which we will now establish. We call μ the permeability of the fluid and μ' that of the wedge. We also denote the angle of the wedge by 2θ. In Fig. 3.3.1.a, $0x$ is the bisector of the wedge. On the upper face of the wedge ($y = x \tan\theta$),

$$H_{\mathrm{n}} = -H_x \sin\theta + H_y \cos\theta$$
$$H_{\mathrm{t}} = H_x \cos\theta + H_y \sin\theta$$

Continuity of the components H_{t} and μH_{n} enables us to compute the components H_x' and H_y' of the magnetic field $\mathbf{H}'$ in the wedge as a function of the components of the magnetic field in region (2), i.e., $\mathbf{H}^{(2)}$:

$$H_x' = H_x^{(2)}\left(\cos^2\theta + \frac{\mu}{\mu'}\sin^2\theta\right) + H_y^{(2)}\sin\theta\cos\theta\left(1 - \frac{\mu}{\mu'}\right)$$
$$H_y' = H_x^{(2)}\sin\theta\cos\theta\left(1 - \frac{\mu}{\mu'}\right) + H_y^{(2)}\left(\sin^2\theta + \cos^2\theta\frac{\mu}{\mu'}\right)$$

Changing θ to $-\theta$, we get the equations connecting the magnetic field $\mathbf{H}'$ with the component of the magnetic field $\mathbf{H}^{(3)}$ in region (3) of the fluid. Elimination of the components H_x' and H_y' then yields a relation between the vectors $\mathbf{H}^{(2)}$ and $\mathbf{H}^{(3)}$,

$$\begin{aligned}\frac{\mu}{\mu'}H_x^{(3)} &= (\alpha\gamma + \beta^2)H_x^{(2)} + 2\beta\gamma H_y^{(2)}\\ \frac{\mu}{\mu'}H_y^{(3)} &= 2\alpha\beta H_x^{(2)} + (\alpha\lambda + \beta^2)H_y^{(2)}\end{aligned} \qquad (3.3.1\text{-}1)$$

with

$$\alpha = \cos^2\theta + \sin^2\theta \frac{\mu}{\mu'}$$

$$\beta = \sin\theta\cos\theta\left(1 - \frac{\mu}{\mu'}\right)$$

$$\gamma = \sin^2\theta + \cos^2\theta \frac{\mu}{\mu'}$$

When we replace the equations $\mathbf{H}^{(2)} = \mathbf{H}^{(3)}$ by the above relations, the number of boundary conditions to be satisfied is unchanged, so that the number of waves is still four. We can write Eqs. (3.3.1-1) in the form

$$\begin{aligned} \mu' H_t^{(3)} &= \cos(2\theta)\,\mu' H_t^{(2)} - \sin(2\theta)\,\mu H_n^{(2)} \\ \mu H_n^{(3)} &= \sin(2\theta)\,\mu' H_t^{(2)} + \cos(2\theta)\,\mu H_n^{(2)} \end{aligned} \tag{3.3.1-2}$$

In practice, the fluid is a paramagnetic medium (i.e., $\mu = 1$), while the wedge is either a paramagnetic medium, $\mu' = 1$, and we have $\mathbf{H}^{(2)} = \mathbf{H}^{(3)}$, or a ferromagnetic medium, $\mu' = \infty$, and we have $H_t^{(2)} = H_t^{(3)} = 0$.

3.3.2. General Remarks on the Piston Problem

The piston problem consists of determining the motion in a fluid conductor of electricity under the action of a piston moving in a prescribed manner. We assume that the fluid is initially at rest and the motion is onedimensional, which means that the unknowns depend on only two variables; the time t and the abcissa x. The difficulty in the case of magnetofluiddynamics is that the continuous motions produced by the piston are not always simple-wave flows. For example, in an expansion, the simple-wave region adjacent to the fluid at rest does not in general extend to the piston. For this reason, we will simplify the problem by considering only the case of a constant-velocity piston.

The fluid motion is defined by the vector $\mathbf{z}$ of components H_x, H_y, H_z, V_x, V_y, V_z, ρ, s, a solution of the equation

$$\frac{\partial \mathbf{z}}{\partial t} + A(\mathbf{z})\frac{\partial \mathbf{z}}{\partial x} = 0 \tag{3.3.2-1}$$

where $A(\mathbf{z})$ is the matrix defined in Section 2.2.3. The piston and the fluid are assumed to have negligible electrical resistivity. Under these conditions, the boundary condition to be satisfied on the piston surface is that the fluid velocity **V** equals the piston velocity **U**. In fact, the slip conditions on the piston surface can be written $V_n - U_n = 0$ (normal velocities equal). In a frame of reference moving with the piston, the electric field $\mathbf{E} = -\mathbf{V} \times \mathbf{B}$ is zero because **V** is zero; hence, the electric field in the fluid adjacent to the piston face is zero (Section 3.1.1, property 9):

$$(\mathbf{V} - \mathbf{U}) \times \mathbf{H} = 0$$

If, as we assume, H_n is not zero, we can deduce that the component of $\mathbf{V} - \mathbf{U}$ on the piston surface is zero; hence, finally, $\mathbf{V} - \mathbf{U} = 0$ on the piston.

Initially, the fluid is homogeneous and at rest: $\mathbf{H} = \mathbf{H}_0$, $\mathbf{V} = 0$, $\rho = \rho_0$, $s = s_0$. In these boundary conditions, t and x only appear in the ratio x/t; on the other hand, Eq. (3.3.2-1) has solutions depending only on the ratio x/t; the motion initiated by the piston is therefore a motion of this type. Such a motion is a simple-wave motion; on the straight line $x = kt$ (k = const), the vector $\mathbf{z}$ stays constant; the simple waves are centered waves.

From the relations satisfied by different wave speeds (Section 3.2.3, Theorems 4 and 5), the solutions of the problem consist simply of one fast wave (shock or expansion wave), one Alfvén wave, and one slow wave (shock or expansion wave); these waves appear in this order and are separated by intervals within which the flow is uniform. If the slow expansion wave is the last to appear, the density of the fluid can fall to zero; a cavity then exists between the piston and the fluid. In this empty cavity, ρ and p are both zero, so that the MFD equations reduce to

$$(\nabla \times \mathbf{H}) \times \mathbf{H} = 0, \qquad \mu \frac{\partial \mathbf{H}}{\partial t} + \nabla \times \mathbf{E} = 0$$

Since the flow is homogeneous (the solution depending only on the ratio x/t), we have, assuming $H_x \neq 0$,

$$\mathbf{H} = \text{const}, \qquad E_y = \text{const}, \qquad E_z = \text{const}$$

On the boundary between the cavity and the fluid, the magnetic field is continuous, since on each surface of discontinuity we have

$$[B_n] = 0, \qquad \rho(V_n - U)[\mathbf{v}] - \mu H_n[\mathbf{h}] = 0$$

The permeability of a vacuum is 1, and that of the fluid usually has the same value; on the other hand $\rho = 0$. We have, therefore,

$$[H_n] = 0, \qquad [\mathbf{h}] = 0$$

The last relation shows that the tangential component of the electric field is continuous; since this is constant in the cavity, its value on the surface ($\mathscr{C}$), which separates the fluid from the cavity, is equal to its value on the piston; and, in a frame of reference moving with the piston, this value is zero. We therefore have on ($\mathscr{C}$)

$$\{(\mathbf{V} - \mathbf{U}) \times \mathbf{H}\}_{\text{tangential}} = 0$$

or

$$(V_z - U_z) H_x - (V_x - U_x) H_z = 0$$
$$(V_x - U_x) H_y - (V_y - U_y) H_x = 0$$

Since H_x is assumed to be nonzero, we deduce that

$$(V_x - U_x)\{(V_y - U_y) H_z - (V_z - U_z) H_y\} = 0$$

Moreover, because the fluid does not extend to the piston, $V_x - U_x \neq 0$, and since the flow is homogeneous, we have the boundary condition

$$(\mathbf{V} - \mathbf{U}) \times \mathbf{H} = 0 \qquad \text{on} \quad (\mathscr{C})$$

i.e., the magnetic field **H** in the cavity is parallel to the difference between the fluid velocity **V** and the piston velocity **U**.

Having established these general results, we now treat two particular piston problems. The first concerns a nonreactive medium, while the second concerns detonations and deflagrations.

3.3.3. Motion of a Piston in a Nonreacting Fluid

The velocity **U** of the piston is arbitrary and the magnetic field in the undisturbed fluid is constant, acting in the x, y plane; the plane of the piston is normal to $0x$

$$\mathbf{U} = \mathbf{x}U_x + \mathbf{y}U_y + \mathbf{z}U_z$$

$$H_0 = \mathbf{x}H_{0x} + \mathbf{y}H_{0y}\,, \qquad H_{0x} > 0, \quad H_{0y} > 0$$

We will work in a space defined by the velocity components V_x, V_y, V_z; the motion of the piston corresponds to a point. We assume that, for any piston velocity, the equations of motion have a unique solution; then, to each point of the velocity space, there corresponds one definite solution of the problem. This solution consists of a certain number of expansion waves and of shock waves, and an Alfvén wave. We seek the loci of points representing motion of the piston in the velocity space. Each locus corresponds to a different group of waves which can originate at the piston. The origin of coordinates in the velocity space corresponds to the undisturbed state of the gas. We first consider flows without cavities.

If a fast expansion wave is propagated in a gas at rest, the points corresponding to the velocity of the fluid behind this wave are located on some curve $0A$ (Fig. 3.3.3.a) along which the following differential

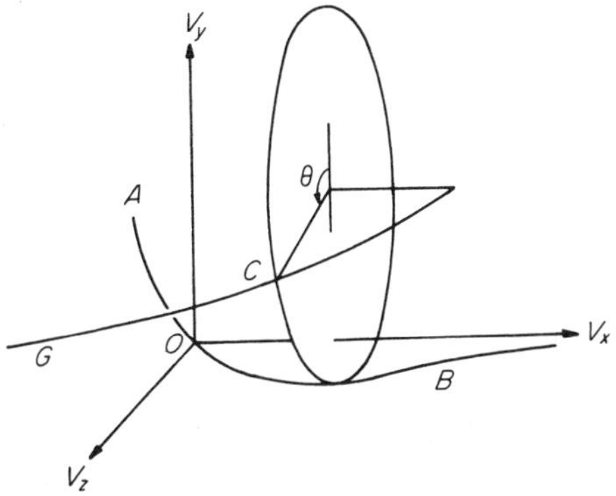

FIG. 3.3.3.a. Fast-wave case.

equations are satisfied:

$$\frac{dH_y}{d\rho} = \frac{H_{0x}}{\rho} s \left[\frac{q-1}{s-(1/q)} \right]^{1/2}$$

$$\frac{dV_x}{d\rho} = \frac{\mu^{1/2} H_{0x}}{\rho \rho^{1/2}} (qs)^{1/2} \qquad (3.3.3\text{-}1)$$

$$\frac{dV_y}{d\rho} = \frac{\mu^{1/2} H_{0x}}{\rho \rho^{1/2}} \left(\frac{q-1}{q-(1/s)} \right)^{1/2}$$

$$H_1 = 0, \qquad V_1 = 0$$

Now, s and q are known functions of ρ,

$$s = \frac{\gamma f(S_0)}{\mu H_{0x}^2} \rho^\gamma \qquad \text{and} \qquad q = q(s)$$

The first equation of (3.3.3-1) can be replaced by the first integral

$$H_y{}^2 = H_{0x}^2(q - 1)[s - (1/q)]$$

The curve $0A$ (along which ρ decreases) is in the part of the plane $V_x < 0$, $V_y > 0$, $V_z = 0$; this curve ends at the point A, where $qs - 1 = 0$, because, beyond A, the magnetic field ceases to be real. At A, the tangent is parallel to the axis of V_y.

If a fast shock wave propagates in the undisturbed gas, the points corresponding to the fluid velocity behind the wave are situated on a curve OB (Fig. 3.3.3.a) on which the shock equations and the inequalities for fast shocks are satisfied; we have, in particular,

$$\begin{aligned} (V_x - D)\rho &= -D\rho_0 \\ (V_x - D)\,\rho V_y - \mu H_{0x} H_y &= -\mu H_{0x} H_{0y} \\ V_z &= 0 \end{aligned}$$

and $\rho > \rho_0$, $H_y > H_{0y}$.

Since the shock speed D is positive, we deduce that $V_x > 0$, $V_y < 0$, and $V_z = 0$, which defines a quadrant of the plane where $0B$ is situated. Hence, when the point representing the piston speed in the velocity space lies on the curve $(S) = 0A + 0B$, the solution of the piston problem contains only one fast wave (expansion or shock).

We assume now that an Alfvén wave is propagated in the fluid with a velocity $\mathbf{x}V_x + \mathbf{y}V_y$. In such a discontinuity, the tangential components of the magnetic fields and of the velocities are related by

$$[\mathbf{v}] = -[\mathbf{h}]\,\mu^{1/2}/\rho^{1/2}$$

Hence, the fluid velocity behind an Alfvén shock can correspond to any point with coordinates

$$V_x, \quad V_y + \frac{\mu H_y}{(\mu\rho)^{1/2}}(1 - \cos\theta), \quad -\frac{\mu H_y}{(\mu\rho)^{1/2}}\sin\theta$$

These points are on a circle of radius $H_y/(\mu\rho)^{1/2}$, having its center at the point

$$V_x, \quad V_y + \frac{\mu H_y}{(\mu\rho)^{1/2}}, \quad 0.$$

The magnetic field behind the Alfvén shock has the components

$$H_{0x}, \quad -H_y \cos\theta, \quad -H_y \sin\theta$$

so that, in the Alfvén shock, the magnetic field is turned through an angle θ. Hence, if the motion of the fluid consists of a fast wave (shock or expansion) and an Alfvén wave, the points corresponding to the piston velocity lie on the surface (Σ) formed by the circles just defined drawn for each point of the curve (S) (Fig. 3.3.3.b).

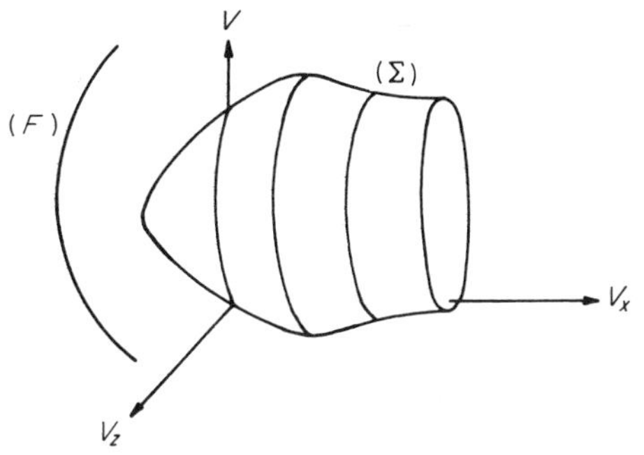

FIG. 3.3.3.b. Fast-wave and Alfvén-wave case.

If the point representing the piston velocity is not on (Σ), the solution of the equation of motion contains a slow wave (shock or expansion). We assume that this is the case and that the fast wave and the Alfvén wave have fixed strength. We call C the point of (Σ) which corresponds to the fast wave and to the Alfvén wave; the points corresponding to the possible piston speeds lie on some curve G passing through C. The vector going from point C to the piston-velocity point represents the velocity of the fluid in the slow wave. since the plane $(\mathbf{n}, \mathbf{H})$ is fixed in a shock as well as in a simple wave, and since the velocity lies in this plane, the whole curve G will lie in the plane passing through this point parallel to the axis $0V_x$ and to the tangential component of magnetic field $\mathbf{y}H_y + \mathbf{z}H_z$; the latter direction does not change in the slow wave. This plane makes an angle θ with the axis $0V_y$, the angle through which the magnetic field turns on crossing the Alfvén wave (Fig. 3.3.3.a). The curves G depend continuously on C and hence cover a certain domain in the velocity space which will be called Q.

In the slow expansion wave, the density of the fluid can vanish while the fluid velocity remains finite; it follows that the loci of the end points of the curves G, corresponding to zero density, form a surface (F) (Fig. 3.3.3.b) which bounds the domain Q. The other end points of G, corresponding to switch-off shocks, lie along a line which is inside the surface (Σ). In fact, in a switch-off shock, we have

$$\rho(V_n - U)[\mathbf{v}] - \mu H_n[\mathbf{h}] = 0$$

$$\rho_1(V_{n_1} - U)^2 - \mu H_n^2 = 0, \qquad \mathbf{h}_2 = 0$$

or

$$\mathbf{v}_2 = \mathbf{v}_1 - \frac{\mathbf{h}_1}{(\mu\rho_1)^{1/2}}$$

At the point C,

$$\mathbf{v}_1 = \mathbf{y}\left\{V_y + \frac{\mu H_y}{(\mu\rho)^{1/2}}(1 - \cos\theta)\right\} - \mathbf{z}\frac{\mu H_y}{(\mu\rho)^{1/2}}\sin\theta$$

$$\mathbf{h}_1 = -\mathbf{y}\,H_y\cos\theta - \mathbf{z}\,H_y\sin\theta$$

We thus have

$$\mathbf{v}_2 = \mathbf{y}\left\{V_y + \frac{\mu H_y}{(\mu\rho)^{1/2}}\right\}$$

value independent of θ. Hence, all the curves G that pass through the points of the same circle converge to some point lying on the line parallel to $0V_x$ through the center of the circle. If the point corresponding to the piston velocity lies in the domain Q, the motion generated by the piston is one in which the fluid stays in contact with the piston; in the contrary case, a cavity develops between the fluid and the piston. The curve G corresponding to the slow wave cannot meet the surface (Σ) except at its initial point, since otherwise two possible flows will correspond to the second point of intersection: one flow which includes a slow wave and one which does not, a result which would lead to a contradiction in the uniqueness hypothesis; the surface (F) therefore does not meet the surface (Σ) and lies outside it. It follows that the flow corresponding to the points of the domain Q lying outside (Σ) are those in which the slow wave is an expansion, while points inside (Σ) correspond to flows in which the slow wave is a shock.

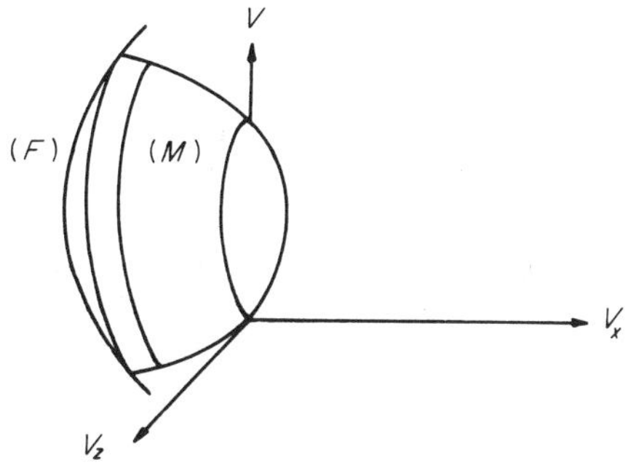

FIG. 3.3.3.c. Alfven-wave and slow-wave case.

The identification of the character of the fast wave can be made in the same fashion. Consider the surface (M) of Fig. 3.3.3.c; this is the locus of piston-velocity points for which the flow contains only an Alfvén wave and a slow wave (shock or expansion). This surface is a surface of revolution with the straight line $V_x = H_{0y}/(\mu\rho_0)^{1/2}$; $V_z = 0$ is its axis of symmetry. If the piston-velocity point is in the domain lying between the surfaces (M) and (F), the fast wave is an expansion wave; if the point is outside this domain but in the region where $\rho > 0$, the fast wave is a shock.

In the case where a cavity is formed between the fluid and the piston, the vector difference $\mathbf{U} - \mathbf{V}$ between the piston speed and the limiting velocity of the fluid is parallel to the vector magnetic field in the cavity. Hence, when the strengths of all waves are fixed, if there is a cavity between the fluid and the piston, the piston-velocity points lie on a line parallel to the cavity magnetic field; this line passes through the point of the surface (F) ($\rho = 0$) representing the limit speed of the fluid.

The diagrams of Figs. 3.3.3.b and 3.3.3.c enable us to determine, for a given speed, the combination of waves which arise in the fluid motion.

3.3.4. Piston Problem for a Reacting Medium

It is interesting to consider the piston problem in the case where the gas is a combustible mixture. One of the two waves (fast or slow) can then be a combustion shock, either a detonation or a deflagration wave in which the gas is burnt. We assume this to be the case, for

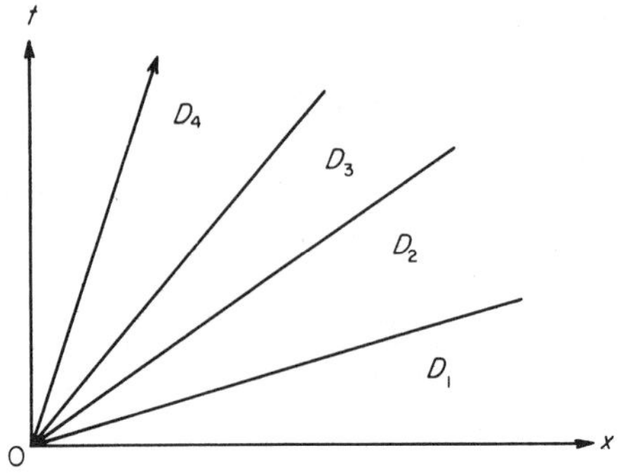

FIG. 3.3.4.a. Domains defined by the three waves, the $0x$ axis, and the piston curve.

instance, for a fast wave (Fig. 3.3.4.a). In the x, t plane, the three waves, the $0x$ axis, and the piston curve define four domains D_1, D_2, D_3, and D_4. The problem is to find in each of these domains a solution of the equation

$$\frac{\partial \mathbf{z}}{\partial t} + A(\mathbf{z}) \frac{\partial \mathbf{z}}{\partial x} = 0 \qquad (3.3.4\text{-}1)$$

where $\mathbf{z}$ is the vector $(H_y, H_z, V_x, V_y, V_z, \rho, s)$ and $A(\mathbf{z})$ the matrix defined in Section 2.2.3. On the $0x$ axis, we must satisfy the initial condition, on the piston the condition $\mathbf{V} = \mathbf{U}$ (the fluid velocity equals piston velocity when the latter is perfect conductor, which we assume is true), and on the shocks the shock equations.

Equation (3.3.4-1) represents a normal hyperbolic system (the eigenvalues of the matrix A are all real); the corresponding boundary conditions are of the mixed Cauchy type. The existence of a solution for this type of problem depends on the orientation of the characteristics for $dt > 0$; it is necessary and sufficient to give on each boundary a number of independent conditions equal to the number of characteristics directed toward the inside of the domain considered.

In the present problem, there are seven characteristics corresponding to the following seven equations:

$$dx/dt = V_x, \qquad dx/dt = V_x + U_A, \qquad dx/dt = V_x - U_A$$
$$dx/dt = V_x + U_-, \qquad dx/dt = V_x - U_-,$$
$$dx/dt = V_x + U_+, \qquad dx/dt = V_x - U_+$$

where U_-, U_+, and U_A are, respectively, the slow and fast magneto-acoustic speeds and the Alfvén speed.

On the $0x$ axis and on the piston curve, the characteristics are oriented as shown in Fig. 3.3.4.b. On $0x$, seven characteristics are

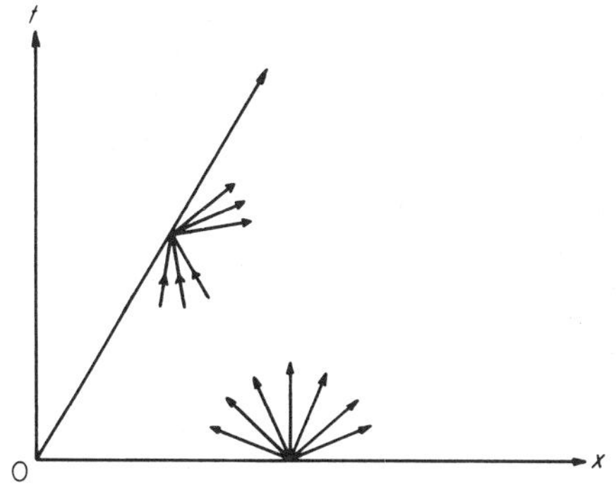

FIG. 3.3.4.b. Orientation of characteristics on the $0x$ axis and piston curve.

directed toward the domain D_1, so that we have to give seven conditions on $0x$, which is the case with the initial values. On the piston curve, only three characteristics are directed toward the domain D_4; we must therefore give three conditions on this boundary, which is the case with the relation $\mathbf{V} = \mathbf{U}$.

The relative directions of the characteristics on each side of the shock (domain D_i in front of the shock and D_{i+1} behind) depend on the shock. Table 3.3.4.I shows this orientation for the eight types of combustion waves which were defined in Section 3.1.6. The shock conditions are

$$[(V_x - D)\,\mathbf{h} - H_x\mathbf{v}] = 0$$

$$[(V_x - D)\,\rho\mathbf{V} + \boldsymbol{\pi}] = 0$$

$$[(V_x - D)\rho] = 0$$

$$(V_x - D)\,\rho\left(\epsilon_i^{(1)} + \frac{V_1^2}{2} + \frac{\mu H_1^2}{\rho_1}\right) + \boldsymbol{\pi}_1 \cdot \mathbf{V}_1 \qquad (3.3.4\text{-}2)$$

$$= (V_x - D)\,\rho\left(\epsilon_i^{(2)} + \frac{V_2^2}{2} + \frac{\mu H_2^2}{\rho_2}\right) + \boldsymbol{\pi}_2 \cdot \mathbf{V}_2$$

TABLE 3.3.4.I

Number of characteristics

	Oriented		Number of conditions to be specified on the shock
	Toward D_i	Toward D_{i+1}	
Strong, fast detonation	0	6	6
Weak, fast deflagration	1	6	7
Strong, fast deflagration	1	7	8
Weak, fast detonation	0	7	7
Strong, slow detonation	2	4	6
Weak, slow detonation	2	5	7
Weak, slow deflagration	3	4	7
Strong, slow deflagration	3	5	8
Nonreacting media			
Slow shock	2	4	6
Fast shock	0	6	6
Alfvén shock	1	5	6

where

$$\pi = (p + \tfrac{1}{2}\mu H^2)\mathbf{x} - \mu H_n \mathbf{H}$$

$$\epsilon_i^{(1)} = f(p_1, \rho_1), \qquad \epsilon_i^{(2)} = g(p_2, \rho_2)$$

The functions f and g are different, but both satisfy the Weyl conditions, and this difference characterizes the combustion of the gas. There are seven conditions, and, since the shock velocity D appears in them, they represent six boundary conditions.

When $\epsilon_i^{(1)}(p, \rho) \equiv \epsilon_i^{(2)}(p, \rho)$, i.e., when there is no combustion, we have only two possibilities: slow shock, fast shock; six characteristics are directed toward the domain D_i or D_{i+1}, so that we have to give six conditions, which is the case with the shock equations. The same is true for the Alfvén shock.

It follows from these considerations that the piston problem is determinate only in the case where the combustion shock is a strong detonation; in the other cases, it will be indeterminate.

The degree of indeterminacy of the piston problem is: 0 for strong detonations, 1 for weak detonations and deflagrations, and 2 for strong deflagrations.

In the case of weak combustion waves, it is necessary to add one more condition so that the problem becomes determinate; this additional condition will be obtained in the next chapter when studying the internal structure of shocks. For strong deflagrations, we need two additional conditions, but the study of shock structure shows that strong deflagrations cannot exist. The results of Section 3.2.4 have already shown that, even if strong deflagrations exist, they will be unstable.

CHAPTER 4

ONE-DIMENSIONAL AND RECTILINEAR FLOWS

Among the solutions of the equations of magnetofluiddynamics, those which depend only on one spatial coordinate can often be written in explicit form. These solutions correspond to the one-dimensional or quasi-one-dimensional flows, the equations of which are derived in the first section of this chapter. The second section is concerned with the particular case of steady one-dimensional flows which are uniform at infinity; this particularly basic case corresponds to the problem of shock-wave structure: the possibility or impossibility for different MFD shocks to be developed as the limit of continuous flow contributes a new criterion of stability. In the third section, nozzle flows are studied; the most remarkable result is the possibility, due to the effect of a magnetic field, of stabilizing decelerating flows in a nozzle. The chapter ends with the study of the electromagnetic flow meter, which is an example of two-dimensional flow in a straight line.

4.1. Equations of Motion

4.1.1. One-Dimensional Flows

The most general equations of a compressible, electrically neutral fluid contain twenty unknowns (see Section 1.2.3), namely, $\mathbf{T}$, $\mathbf{H}$, $\mathbf{E}$, $\mathbf{V}$, $\boldsymbol{\theta}$, ρ, and ϵ. The pressure and the temperature are known functions of the density ρ and the specific internal energy ϵ_i; in the case of a polytropic gas, we have

$$p = (\gamma - 1)\,\rho\epsilon_i\,, \qquad T = (1/C_v)\,\epsilon_i$$

The equations contain four dissipation coefficients, the two coefficients of viscosity λ_1 and μ_1, the coefficient of thermal conductivity λ, and the coefficient of electrical conductivity σ. The equations have solutions in which the unknowns depend only on the time and one space variable (which we shall take as the abscissa x); these one-dimensional flows are governed by the following simplified equations of motion:

$$\rho\left(\frac{\partial \mathbf{V}}{\partial t} + u\,\frac{\partial \mathbf{V}}{\partial x}\right) = \nabla \cdot \mathbf{T} + \mathbf{J} \times \mu\mathbf{H}$$

$$\frac{\partial \rho}{\partial t} + \frac{\partial(\rho u)}{\partial x} = 0$$

$$\frac{\partial \Lambda}{\partial t} + \frac{\partial(\Lambda u)}{\partial x} = \nabla \cdot (\mathbf{T} \cdot \mathbf{V}) - \frac{\partial \Theta}{\partial x} + \mathbf{J} \cdot \mathbf{E}$$

$$\frac{\partial \mu H_x}{\partial t} = 0, \qquad \frac{\partial \mu H_y}{\partial t} = \frac{\partial E_z}{\partial x}, \qquad \frac{\partial \mu H_z}{\partial t} = -\frac{\partial E_y}{\partial x} \tag{4.1.1-1}$$

$$J_x = 0, \qquad J_y = -\frac{\partial H_z}{\partial x}, \qquad J_z = \frac{\partial H_y}{\partial x}$$

$$\mathbf{T} = \left(-p + \lambda_1 \frac{\partial u}{\partial x}\right)\mathbf{U} + \mu_1 \begin{Vmatrix} 2\,\partial u/\partial x & \partial v/\partial x & \partial w/\partial x \\ \partial v/\partial x & 0 & 0 \\ \partial w/\partial x & 0 & 0 \end{Vmatrix}$$

$$\Theta = -\lambda \frac{\partial T}{\partial x}$$

In these equations, u, v, and w denote the components of the fluid velocity $\mathbf{V}$, and $\Lambda = \rho(\epsilon_i + \frac{1}{2} V^2)$, the specific total energy. To these equations, we must add

$$\nabla \cdot \mathbf{H} = 0 \qquad \text{and} \qquad \mathbf{J} = \sigma(\mathbf{E} + \mathbf{V} \times \mu\mathbf{H})$$

The component H_x is then constant; moreover, the vector $\boldsymbol{\Theta}$ has only one nonzero component, and the tensor $\mathbf{T}$ only four different nonzero components. Equations (4.1.1-1) then form a system of 15 equations in 15 unknowns.

We shall use the following results:

$$\nabla \cdot \mathbf{T} = \left| \begin{array}{l} \dfrac{\partial}{\partial x}\left\{-p + (\lambda_1 + 2\mu_1)\dfrac{\partial u}{\partial x}\right\} \\[2ex] \dfrac{\partial}{\partial x}\left\{\mu_1 \dfrac{\partial v}{\partial x}\right\} \\[2ex] \dfrac{\partial}{\partial x}\left\{\mu_1 \dfrac{\partial w}{\partial x}\right\} \end{array} \right.$$

$$\mathbf{J} \times \mu\mathbf{H} = \left| \begin{array}{l} -\dfrac{\partial}{\partial x}\left(\mu \dfrac{H_y^2 + H_z^2}{2}\right) \\[2ex] \mu H_x \dfrac{\partial H_y}{\partial x} \\[2ex] \mu H_x \dfrac{\partial H_z}{\partial x} \end{array} \right.$$

$$\nabla \cdot (\mathbf{T} \cdot \mathbf{V}) = \frac{\partial}{\partial x}\left\{\left(-p + (\lambda_1 + 2\mu_1)\frac{\partial u}{\partial x}\right) u + \mu_1 \frac{\partial v}{\partial x} v + \mu_1 \frac{\partial w}{\partial x} w\right\}$$

a. Case of Steady Motion

In the case of steady motion, Eqs. (4.1.1-1) simplify considerably and, in particular, have an important number of first integrals. We then obtain the following equations:

$$\begin{aligned} \rho u &= m \\ mu &= -p + (\lambda_1 + 2\mu_1)\frac{du}{dx} - \frac{\mu(H_y^2 + H_z^2)}{2} + P \\ mv &= \mu_1 \frac{dv}{dx} + \mu H_x H_y + P_1 \\ mw &= \mu_1 \frac{dw}{dx} + \mu H_x H_z + P_2 \end{aligned}$$

$$m\left(\epsilon_i + \frac{u^2 + v^2 + w^2}{2}\right) = \left(-p + (\lambda_1 + 2\mu_1)\frac{du}{dx}\right) u + \mu_1 \frac{dv}{dx} v + \mu_1 \frac{dw}{dx} w + \lambda \frac{dT}{dx} - E_y H_z + E_z H_y + C$$

$$E_y = \text{const}, \qquad E_z = \text{const}$$

$$0 = E_x + v\mu H_z - w\mu H_y$$

$$\sigma^{-1}\frac{dH_y}{dx} = E_z + u\mu H_y - v\mu H_x$$

$$\sigma^{-1}\frac{dH_z}{dx} = -E_y + u\mu H_z - w\mu H_x$$

where m, P, P_1, P_2, and C denote constants of integration. The investigation of steady motion is thus reduced to the solution of a system of six equations in six unknowns. The unknowns are H_y, H_z, u, v, w, and T; ρ is equal to m/u, p and ϵ_i are known functions of ρ and T. We write the equations in the following form:

$$\begin{aligned}
(\lambda_1 + 2\mu_1)\frac{du}{dx} &= mu + p + \frac{\mu(H_y^2 + H_z^2)}{2} - P \\
\mu_1 \frac{dv}{dx} &= mv - \mu H_x H_y - P_1 \\
\mu_1 \frac{dw}{dx} &= mw - \mu H_x H_z - P_2 \\
\lambda \frac{dT}{dx} &= m\left(\epsilon_i - \frac{u^2 + v^2 + w^2}{2}\right) - u\mu\frac{H_y^2 + H_z^2}{2} \\
&\quad + \mu H_x(vH_y + wH_z) + uP + vP_1 + wP_2 \\
&\quad + E_y H_y + E_z H_z - C \\
\sigma^{-1}\frac{dH_y}{dx} &= E_z + u\mu H_y - v\mu H_x \\
\sigma^{-1}\frac{dH_z}{dx} &= -E_y + u\mu H_z - w\mu H_x
\end{aligned} \tag{4.1.1-2}$$

We thus arrive at a system of six differential equations of first order for the six unknowns, which are solved here for the first derivatives. These equations contain four dissipation coefficients, $\lambda_1 + 2\mu_1$, μ_1, λ, and σ^{-1}. When certain of these coefficients are zero, the system is reduced to r differential equations and $6 - r$ algebraic equations. The variables which have derivatives appearing in the system are called primary, the others are called secondary.

The equations are further simplified if we assume that motion is in a straight line ($v = w = 0$).

b. Case of Steady Rectilinear Motion

Since the components E_y and E_z of the electric field are constant, we can always find a frame of reference in which $E_y = 0$. Then E_z is the only nonzero component, since $v = w = 0$ implies $E_x = 0$. Putting $E_z = E$, the last of Eqs. (4.1.1-2) gives

$$H_z(x) = H_z(-\infty) \exp \int_{-\infty}^{x} \sigma\mu \, u(x) \, dx$$

Conditions at infinity normally require a nonzero velocity and a bounded magnetic field there. It follows that we have $H_z(-\infty) = 0$; then $H_z(x) \equiv 0$. The second of Eqs. (4.1.1-2) shows that the product $H_x H_y$ is constant; since H_x is constant, the magnetic field will be different from a constant if $H_x = 0$; then $P_1 = 0$, as we shall assume. Finally, the electromagnetic field vectors have the following components:

$$\mathbf{H} = \begin{vmatrix} 0 \\ H(x) \\ 0 \end{vmatrix}, \qquad \mathbf{E} = \begin{vmatrix} 0 \\ 0 \\ E \end{vmatrix}$$

The three unknown functions H, u, and T are solutions of the following system:

$$\begin{aligned} (\lambda_1 + 2\mu_1) \frac{du}{dx} &= mu + p + \frac{\mu H^2}{2} - P \\ \lambda \frac{dT}{dx} &= m\left(\epsilon_i - \frac{u^2}{2}\right) - u \frac{\mu H^2}{2} + uP + EH - C \qquad (4.1.1\text{-}3) \\ \sigma^{-1} \frac{dH}{dx} &= u\mu H + E \end{aligned}$$

For a polytropic gas, we must add

$$p = m \frac{RT}{u}, \qquad \epsilon_i = C_V T$$

4.1.2. Quasi-One-Dimensional Flows

We say that a flow is quasi-one-dimensional when it is possible to neglect all variations of the characteristic quantities in a direction normal to a given direction, taken along the axis $0x$; the same condition applies to mechanical as well as to electromagnetic quantities.

Such an approximation is justified in the study of flow in a nozzle with an axis of symmetry (or an axis of revolution) and a slowly varying cross section. We shall denote by Q the cross section of the nozzle and we write the equations of motion neglecting the small quantities v, w, H_x, E_x, E_y and their derivatives in the x direction. The momentum equations then become

$$\rho\left(\frac{\partial u}{\partial t} + u\frac{\partial u}{\partial x}\right) + \frac{\partial p}{\partial x} = -J_z \mu H_y$$
$$\frac{\partial p}{\partial y} = 0, \qquad \frac{\partial p}{\partial z} = 0 \tag{4.1.2-1}$$

The law of conservation of energy gives

$$\frac{\partial S}{\partial t} + u\frac{\partial S}{\partial x} = \sigma^{-1} J_z^{\,2} \tag{4.1.2-2}$$

$$J_z = \sigma(E_z + u\mu H_y) \qquad \text{(Ohm's law)} \tag{4.1.2-3}$$

The derivatives $\partial v/\partial y$ and $\partial w/\partial z$ appear in the continuity equation, and these cannot be neglected *a priori*. Expressing the fact that the mass of fluid contained between two planes with abscissae $a(t)$ and $b(t)$, moving with the fluid, is constant, we find

$$\frac{d}{dt}\int_a^b \rho Q\, dx = 0$$
$$(u\rho Q)_{x=b} - (u\rho Q)_{x=a} + \int_a^b \frac{\partial(\rho Q)}{\partial t}\, dx = 0 \tag{4.1.2-4}$$
$$\frac{\partial(\rho Q)}{\partial t} + \frac{\partial(u\rho Q)}{\partial x} = 0$$

We must of course add Ohm's law to these equations. Thus, to determine the unknowns **E**, **H**, **V**, p, and ρ, we must add to the above equations two additional vector equations.

As a first approximation, we regard the electromagnetic field **E**, **H** as being given by an external process and neglect electromagnetic effects induced by the motion; the problem then contains only the unknowns **V**, p, and ρ.

If we wish to take account of induced fields, we must write down the two Maxwell equations, which can also be written in a quasi-one-dimensional form. The local values **e**, **h**, and **j** of the fields and electric current, variables at a cross section, satisfy Faraday's law and Ampère's theorem:

$$\frac{d}{dt}\iint_{(S)} \mu\mathbf{h}\cdot\mathbf{n}\,ds + \int_{(C)} \mathbf{E}\cdot d\mathbf{l} = 0$$

$$\iint_{(S)} \mathbf{j}\cdot\mathbf{n}\,ds - \int_{(C)} \mathbf{h}\cdot d\mathbf{l} = 0 \tag{4.1.2-5}$$

where (C) denotes a certain closed curve and (S) a surface bounded by this curve with unit normal vector $\mathbf{n}$. The variations of $\mathbf{e}$, $\mathbf{h}$, $\mathbf{j}$ over a cross section being supposed small, we can, as a first approximation, replace them by their mean values over this section, say, $\mathbf{E}$, $\mathbf{H}$, and $\mathbf{J}$. The form obtained for Maxwell's equations depends on the nozzle and the direction of the electric and magnetic fields. We shall consider three main cases (Fig. 4.1.2.a). Writing in each of these cases Faraday's

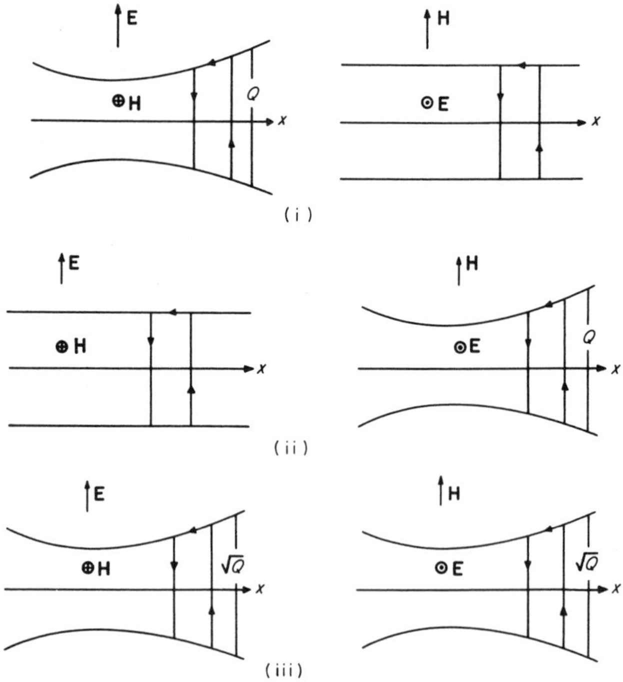

FIG. 4.1.2.a. Faraday's law (left) and Ampère's theorem (right) for: (i) two-dimensional nozzle normal to magnetic field; (ii) two-dimensional nozzle normal to the electric field; (iii) nozzle with homothetic cross section.

law on the contour shown in the figure on the left, and Ampère's theorem on the contour shown on the right, we obtain Maxwell's equations in the following form:

(i) Two-dimensional nozzle perpendicular to the magnetic field:

$$\frac{\partial \mu H}{\partial t} - \frac{1}{Q}\frac{\partial (EQ)}{\partial x} = 0, \qquad \frac{\partial H}{\partial x} - J = 0$$

(ii) Two-dimensional nozzle normal to the electric field:

$$\frac{\partial \mu H}{\partial t} - \frac{\partial E}{\partial x} = 0, \qquad \frac{1}{Q}\frac{\partial (HQ)}{\partial x} - J = 0$$

(iii) Nozzle with homothetic cross section:

$$\frac{\partial \mu H}{\partial t} - \frac{1}{\sqrt{Q}}\frac{\partial}{\partial x}(E\sqrt{Q}) = 0, \qquad \frac{1}{\sqrt{Q}}\frac{\partial}{\partial x}(H\sqrt{Q}) - J = 0$$

The three cases can be reduced to the two equations

$$\frac{\partial \mu H}{\partial t} - \frac{1}{Q^\alpha}\frac{\partial}{\partial x}(EQ^\alpha) = 0$$
$$\frac{1}{Q^\beta}\frac{\partial}{\partial x}(HQ^\beta) - J = 0, \qquad \alpha + \beta = 1 \tag{4.1.2-6}$$

to which must be added Ohm's law

$$J = \sigma(E + u\mu H) \tag{4.1.2-7}$$

For a nozzle plane perpendicular to **H**, $\alpha = 1$; for a nozzle plane perpendicular to **E**, $\alpha = 0$; for a nozzle with homothetic cross section, $\alpha = \frac{1}{2}$.

4.2. Structure of Shock Waves

4.2.1. General Study of Shock Layers

The study of discontinuities given in the previous chapter was of those appearing in flows of perfect fluids. These shocks, in fact, represent a more complex physical phenomenon and the discontinuities really represent the limit of flow in a domain of very small thickness in which there exists some sort of equilibrium between mechanical nonlinear effects and opposing dissipation effects. To study the structure of shock waves, we make a local study of these continuous flows and determine their limit when the coefficients of dissipation tend to zero. Because it is a local study in time and space, we can limit ourselves to the investigation of one-dimensional steady flows.

The equations of motion, (4.1.1-2), can be written in the form

$$\epsilon_i \frac{dz_i}{dx} = F_i(z_1, z_2, \ldots, z_n), \qquad i = 1, 2, \ldots, n \tag{4.2.1-1}$$

Such a system, in which the terms on the right, F_i, are independent of the variable x, is called an autonomous system. The coefficients ϵ_i that can depend on $\mathbf{z}(z_1, z_2, \ldots, z_r)$, but not on x, are the coefficients of dissipation, and some of them may be zero. The problem of general study of shock layers consists in finding a solution of the system (4.2.1-1) which is uniform both at $x = -\infty$ and at $x = +\infty$. At these points, the derivatives dz_i/dx must be zero; hence, in Euclidean space, the point $z_1, z_2, \ldots, z_n$ corresponding to those at infinity (upstream and downstream) are points of the surfaces $F_i(z_1, z_2, \ldots, z_n) = 0$; they are the singular points of the system (4.2.1-1). Only one integral curve can pass through a regular point, while several curves can pass through a singular point.

Once the singular points $S_1, S_2, \ldots$ have been determined, the study of the integral curves in the neighborhood of this point can be made by linearized methods. Denoting the value of $\mathbf{z}$ at a singular point by $\mathbf{z}^*$, the variation of the function $\mathbf{z}(x)$ near this point will be given by

$$\epsilon_i(\mathbf{z}^*) \frac{d}{dx}(z_i - z_i^*) = \nabla F_i(z^*) \cdot (\mathbf{z} - \mathbf{z}^*)$$

or

$$\frac{d(\delta\mathbf{z})}{dx} = A(\mathbf{z}^*)\, \delta\mathbf{z} \tag{4.2.1-2}$$

where $\delta\mathbf{z} = \mathbf{z} - \mathbf{z}^*$, while the matrix $A(\mathbf{z})$ is formed from the vectors $\nabla F_i(\mathbf{z})$ and the scalars $\epsilon_i(\mathbf{z}^*)$. The general solution of the system (4.2.1-2) is

$$\delta\mathbf{z} = \sum_{i=1}^{n} c_i \mathbf{r}_i(\mathbf{z}^*) \exp[g_i(\mathbf{z}^*)x] \tag{4.2.1-3}$$

where the $g_i(\mathbf{z}^*)$ are the eigenvalues of the matrix $A(\mathbf{z}^*)$, the $\mathbf{r}_i(\mathbf{z}^*)$ are the eigenvectors corresponding to $\lambda_i(\mathbf{z}^*)$, and the c_i are arbitrary constants.

For a singular point $\mathbf{z}^*$ to represent infinity upstream ($x = -\infty$), it is necessary that at least one of the eigenvalues g_i be positive. If it is to represent downstream infinity ($x = +\infty$) one of the eigenvalues g_i must be negative.

The problem of general study of shock layers then reduces to finding the integrals of the system (4.2.1-1) which join a singular point $\mathbf{z}^*$ representing infinity upstream to one $\mathbf{z}^{*\prime}$ at infinity downstream. Let r be the number of positive eigenvalues and r' be the number of negative eigenvalues of the matrices $A(\mathbf{z}^*)$ and $A(\mathbf{z}^{*\prime})$, respectively. The integral curves starting from the point $\mathbf{z}^*$ depend on r parameters, those constants c_i corresponding to the positive eigenvalues (the other c_i must be zero). In fact, these constants are defined to within a factor which corresponds to a translation of the integral curve. To state that an integral curve passes through the point $\mathbf{z}^{*\prime}$, we have to write $n - r'$ conditions knowing that the constants c_i corresponding to positive eigenvalues of $A(\mathbf{z}^{*\prime})$ are zero. To state that an integral curve goes from $\mathbf{z}^*$ at $x = -\infty$ to $\mathbf{z}^{*\prime}$ at $x = \infty$, we must impose $n - r'$ conditions homogeneous in r parameters defined to within a constant factor. Three cases can arise: $n - r' > r - 1$, the problem is impossible; $n - r' = r - 1$, there exist a finite number of solutions; $n - r' < r - 1$, the problem is underdetermined.

The number $n + 1 - r - r'$ is called the degree of overdetermination of the problem.

When this degree is zero, we can look for integral curves starting from one of the singular points. It is then important to know the position taken by the eigenvectors with respect to the regions of the space $z_1, z_2, \ldots, z_n$ determined by the surfaces $F_i(z_1, \ldots, z_n) = 0$. The study is simple when one of the two numbers r and r' is equal to 1 and consequently the other equal to n. We will next study the shape of the integral curves when they meet the surfaces $F_i = 0$ and see if it is possible to reach the other singular point.

Before applying these general remarks to the MFD case, we will study in more detail the existence and the uniqueness of solutions of shock-layer problems when $n = 2$.

4.2.2. Existence and Uniqueness of Shock Layers

When the vector $\mathbf{z}$ has only two components, the system of equations (4.2.1-1) becomes

$$\begin{aligned} \epsilon_1\, dz_1/dx &= F_1(z_1, z_2) \\ \epsilon_2\, dz_2/dx &= F_2(z_1, z_2) \end{aligned} \qquad (4.2.2\text{-}1)$$

where the coefficients $\epsilon_1(z_1, z_2)$ and $\epsilon_2(z_1, z_2)$ are given positive functions, while the functions $F_1(z_1, z_2)$ and $F_2(z_1, z_2)$ satisfy the following hypotheses:

($\mathfrak{H}_1$) $\partial F_1/\partial z_2 > 0$, $\partial F_2/\partial z_2 > 0$.

($\mathfrak{H}_2$) The curves (F_1) and (F_2) with equations $F_1(z_1, z_2) = 0$ and $F_2(z_1, z_2) = 0$ intersect in two and only two points $\mathbf{z}^*(z_1^*, z_2^*)$ and $\mathbf{z}^{*\prime}(z_1^{*\prime}, z_2^{*\prime})$. We will choose $z_1^* > z_1^{*\prime}$.

($\mathfrak{H}_3$) $\partial F_2/\partial z_1 > 0$ on (F_2) for $z_1^{*\prime} < z_1 < z_1^*$.

($\mathfrak{H}_4$)

$$\begin{vmatrix} \partial F_1/\partial z_1 & \partial F_1/\partial z_2 \\ \partial F_2/\partial z_1 & \partial F_2/\partial z_2 \end{vmatrix} > 0 \quad \text{at} \quad \mathbf{z}^* \quad \text{and} \quad < 0 \quad \text{at} \quad \mathbf{z}^{*\prime}$$

We will note, to start with, that, as a consequence of ($\mathfrak{H}_1$), the curves (F_1) and (F_2) can be represented by uniform functions of z_1: $z_2 = f_1(z_1)$ and $z_2 = f_2(z_1)$, respectively. According to hypothesis ($\mathfrak{H}_3$), the function $f_2(z_1)$ is monotonic and decreasing in the interval $z_1^{*\prime} < z_1 < z_1^*$, and consequently $z_2^{*\prime} > z_2$.

According to hypothesis ($\mathfrak{H}_2$) and ($\mathfrak{H}_4$), we have, in the interval $z_1^{*\prime} < z_1 < z_1^*$: $f_2(z_1) < f_1(z_1)$. The closed curve formed by the arcs (F_1^*) of (F_1) and (F_2^*) of (F_2), joining the points $\mathbf{z}^*$ and $\mathbf{z}^{*\prime}$, is the boundary of a simply connected domain $(\mathscr{R})$ of the plane (z_1, z_2). We then conclude from hypothesis ($\mathfrak{H}_1$), that in the whole domain $(\mathscr{R})$, we have

$$F_1(z_1, z_2) < 0 \qquad \text{and} \qquad F_2(z_1, z_2) > 0$$

The linear equations associated with the system (4.2.2-1) are

$$\frac{d(\delta \mathbf{z})}{dx} = A(\mathbf{z})\, \delta \mathbf{z}$$
$$A(\mathbf{z}) = \begin{pmatrix} (1/\epsilon_1)\, \partial F_1/\partial z_1 & (1/\epsilon_1)\, \partial F_1/\partial z_2 \\ (1/\epsilon_2)\, \partial F_2/\partial z_1 & (1/\epsilon_2)\, \partial F_2/\partial z_2 \end{pmatrix} \tag{4.2.2-2}$$

The characteristic equation of the matrix $A(\mathbf{z})$ can be written

$$g^2 - \left(\frac{1}{\epsilon_1}\frac{\partial F_1}{\partial z_1} + \frac{1}{\epsilon_2}\frac{\partial F_2}{\partial z_2}\right) g + \frac{1}{\epsilon_1 \epsilon_2}\begin{vmatrix} \partial F_1/\partial z_1 & \partial F_1/\partial z_2 \\ \partial F_2/\partial z_1 & \partial F_2/\partial z_2 \end{vmatrix} = 0$$

At the point $\mathbf{z}^*$, the two eigenvalues are real and positive; at the point $\mathbf{z}^{*\prime}$, they are real and of opposite sign; the point $\mathbf{z}^*$ then represents infinity upstream ($x = -\infty$); the degree of overdetermination is zero.

A solution of the system (4.2.2-1), $\mathbf{z}(x)$, corresponding to given functions $\epsilon_1(z_1, z_2)$ and $\epsilon_2(z_1, z_2)$, will be called a shock layer if

$$\lim_{x \to -\infty} \mathbf{z}(x) = \mathbf{z}^* \qquad \text{and} \qquad \lim_{x \to \infty} \mathbf{z}(x) = \mathbf{z}^{*\prime}$$

The corresponding shock curve in the (z_1, z_2) plane is the integral curve represented by one of the solutions $\mathbf{z}(x + h)$, where h is an arbitrary constant. All the shock layers corresponding to different values of h will be considered identical, and the shock layer will be said to be parametrized once a value of h is chosen.

Theorem 1. For any pair of functions $\epsilon_1(z_1, z_2)$ and $\epsilon_2(z_1, z_2)$, there exists one and only one shock layer joining the points $\mathbf{z}^*$ and $\mathbf{z}^{*\prime}$.

Only one eigenvalue of the matrix $A(\mathbf{z})$ is negative at the point $\mathbf{z}^{*\prime}$; there then exist two and only two integrals that reach this point for $x = +\infty$. The two curves have the same tangent at the point $\mathbf{z}^{*\prime}$, but reach this point from opposite directions. The tangent is given by the eigenvector corresponding to the negative eigenvalue of

$$\frac{dz_1}{(1/\epsilon_1)(\partial F_1/\partial z_2)} = \frac{dz_2}{g - (1/\epsilon_1)(\partial F_1/\partial z_1)}$$

As g is negative, dz_1 and dz_2 will have opposite signs. The curves (F_1) and (F_2) determine four regions in the neighborhood of the point $\mathbf{z}^{*\prime}$, and the integral curves that reach this point for $x = \infty$ are situated in the regions where dz_1 and dz_2 (or F_1 and F_2) have opposite signs. The region $(\mathscr{R})$ is one of those, the other being the opposite region. In $(\mathscr{R})$, we have, in fact, $F_1 < 0$ and $F_2 > 0$. We call $\mathbf{z}(x)$ the integral curve that reaches $\mathbf{z}^{*\prime}$ in the region $(\mathscr{R})$ and we will prove that $\mathbf{z}(x)$ is a shock layer (Fig. 4.2.2.a). For this, we will consider integral curves of the system (4.2.2-1) which pass through points of (F_1^*) and (F_2^*). On (F_1^*), we have $F_1 = 0$ and $F_2 > 0$, which means that the integral curves have a tangent parallel to $0z_2$ and are directed toward the outside when x increases. On (F_2^*), they have a tangent parallel to $0z_1$ and are also directed toward the outside when x increases. Hence, for x decreasing, the curve $\mathbf{z}(x)$ cannot leave the region $(\mathscr{R})$ and has a negative slope. Because it can only terminate in a singular point, it reaches the point $\mathbf{z}^*$ for $x \to -\infty$, which proves the existence of a shock layer.

To prove uniqueness, we will show that no other integral curve can

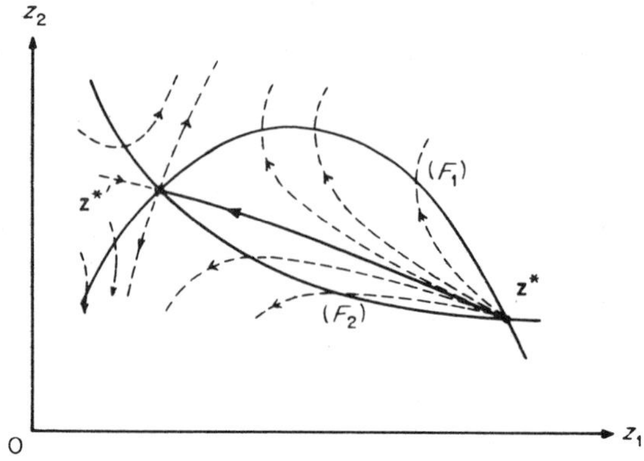

FIG. 4.2.2.a. Shock curve.

connect $\mathbf{z}^*$ and $\mathbf{z}^{*\prime}$. Since the point $\mathbf{z}^*$ can only be reached when $x \to -\infty$, then the only solution other than $\mathbf{z}(x)$ will be the second integral curve $\mathbf{z}(x)$ which reaches $\mathbf{z}^{*\prime}$ for $x = +\infty$. This is impossible because the two shock layers will form a closed curve bounding a simply connected region of the plane (z_1, z_2). One of the integral curves starting from $\mathbf{z}^{*\prime}$ at $x = -\infty$ enters this domain, and cannot reach $\mathbf{z}^*$ (which never corresponds to $x = \infty$), nor leave this domain, because it would intersect the boundary, which is an integral curve, in a regular point (and only one integral curve goes through a regular point). Further, it cannot have a limit cycle, because inside the limit cycle there is a singular point. These contradictions prove the uniqueness of the shock layer.

We will now study the behavior of the shock layer when $\epsilon_1(z_1, z_2)$ and $\epsilon_2(z_1, z_2)$ tend to zero. For given functions ϵ_1 and ϵ_2, we will call $\mathbf{z}(x; \epsilon_1, \epsilon_2)$ the shock layer and $S(\epsilon_1, \epsilon_2)$ the associated curve of the (z_1, z_2) plane. We assume that ϵ_1 and ϵ_2 can be made arbitrarily small in $(\mathscr{R})$, and in the following all bounds will be in the region $(\mathscr{R})$. The shock layers will be assumed to be conveniently parametrized.

Theorem 2. Let $\mathbf{z} = \mathbf{z}^*$ for $x < \xi$ and $\mathbf{z} = \mathbf{z}^{*\prime}$ for $x > \xi$ be a shock wave The family of shock layers $\mathbf{z}(x; \epsilon_1, \epsilon_2)$, conveniently parameterized, converges to the above shock wave when $\epsilon_1 \to 0$ and $\epsilon_2 \to 0$. The convergence is uniform in any closed interval not containing the point $x = \xi$.

To prove this new theorem, take a positive number η and call

$\mathscr{R}(\eta)$ the subregion of $(\mathscr{R})$ exterior to the circles of centers $\mathbf{z}^*$ and $\mathbf{z}^{*\prime}$ with radius η. By Eq. (4.2.2-1), we have, inside $\mathscr{R}(\eta)$,

$$\frac{d(z_2 - z_1)}{dx} = -\frac{F_1(z_1, z_2)}{\epsilon_1(z_1, z_2)} + \frac{F_2(z_1, z_2)}{\epsilon_2(z_1, z_2)} > 0$$

Define $\epsilon = \sup(\epsilon_1, \epsilon_2)$ in $(\mathscr{R})$ and let $C(\eta)$ be a constant such that

$$|F_1(z_1, z_2)| + F_2(z_1, z_2) \geqslant C(\eta) > 0 \qquad \text{within} \quad \mathscr{R}(\eta)$$

Such a constant exists. For every integral curve, we have

$$\frac{d(z_2 - z_1)}{dx} > \frac{C(\eta)}{\epsilon} > 0 \qquad \text{in} \quad \mathscr{R}(\eta)$$

We then consider a shock layer $\mathbf{z}(x; \epsilon_1, \epsilon_2)$ and the intersections $\mathbf{z}_m$ and $\mathbf{z}_M$ of this layer with the circles of radius η centered at $\mathbf{z}^*$ and

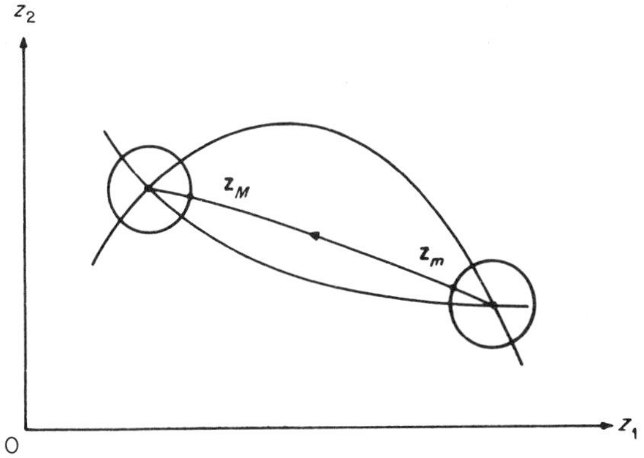

FIG. 4.2.2.b. Intersections z_m and z_M of shock layer with circles of radius η.

$\mathbf{z}^{*\prime}$ (Fig. 4.2.2.b). We have $\mathbf{z}(x_m; \epsilon_1, \epsilon_2) = \mathbf{z}_m$ and $\mathbf{z}(x_M; \epsilon_1, \epsilon_2) = \mathbf{z}_M$. For $x_m \leqslant x \leqslant x_M$, we have

$$z_1^* > z_{1m} \geqslant z_1(x; \epsilon_1, \epsilon_2) \geqslant z_{1M} > z_1^{*\prime}$$

$$z_2^* < z_{2m} \leqslant z_2(x; \epsilon_1, \epsilon_2) \leqslant z_{2M} < z_2^{*\prime}$$

From the inequality concerning the derivative $d(z_2 - z_1)/dx$, we get

$$x_M - x_m \leqslant \frac{\epsilon}{C(\eta)} \{(z_{2M} - z_{1M}) - (z_{2m} - z_{1m})\}$$

$$x_M - x_m < \epsilon \frac{z_2^{*\prime} - z_2^* + z_1^* - z_1^{*\prime}}{C(\eta)} < \delta$$

Hence, given η and δ, we can always find ϵ such that, when going from $\mathbf{z}_m$ to $\mathbf{z}_M$, x will vary by as little as we please. We can always parameterize the shock layer so that the interval x_m, x_M includes the value $x = \xi$.

We will finish by finding the limit of the shock layer when only one of the coefficients ϵ_1, ϵ_2 tends to zero, the other being a known function of z_1, z_2.

Theorem 3. Given an open neighborhood G of the closed arc $(F_2{}^*)$, every shock curve $S(\epsilon_1, \epsilon_2)$ will lie entirely in the domain G as long as the ratio ϵ_2/ϵ_1 is small enough (Fig. 4.2.2.c).

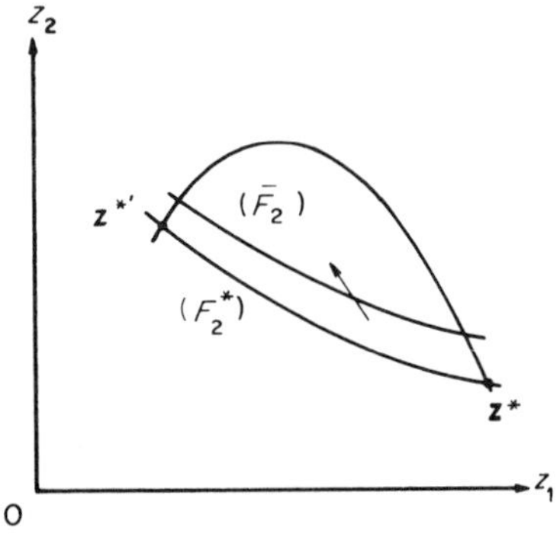

FIG. 4.2.2.c. Situation for ϵ_2/ϵ_1 small.

We will call $(\bar{F}_2)$ an arc with bounded negative slope which ends on $(F_1{}^*)$ but not at $\mathbf{z}^*$ or $\mathbf{z}^{*\prime}$. We will assume $(\bar{F}_2)$ to be close enough to $(\bar{F}_2{}^*)$ so that all the points of $(\mathscr{R})$ between $(\bar{F}_2)$ and $(F_2{}^*)$ belong to G. Such a curve $(\bar{F}_2)$ can always be found, because $(F_2{}^*)$ is monotonic. We will call D the complement of G in $(\mathscr{R})$. On $(\bar{F}_2)$, we have

$$F_2(z_1, z_2) \geqslant k_1 > 0, \qquad |F_1(z_1, z_2)| < k_2, \qquad |\text{slope of } (\bar{F}_2)| \leqslant N$$

where k_1, k_2, and N are suitable positive constants. We assume that we take $\epsilon_1/\epsilon_2 > (k_1/k_2)\,N$. On $(\bar{F}_2)$, the slopes of the integral curves satisfy the relations

$$-\frac{dz_2}{dz_1} = -\frac{\epsilon_1}{\epsilon_2}\frac{F_2(z_1, z_2)}{F_1(z_1, z_2)} \geqslant \frac{\epsilon_1}{\epsilon_2}\frac{k_1}{k_2} > N$$

These integral curves are directed toward D (for increasing x).

Consequently, an integral curve containing a point of D cannot intersect $(\bar{F}_2)$ again (for x increasing), and consequently cannot reach $\mathbf{z}^{*\prime}$. The shock layers for which the ratio ϵ_2/ϵ_1 is small enough are then entirely in the region G.

Analogous considerations apply to the case where ϵ_1/ϵ_2 is small (Fig. 4.2.2.d), but new results arise from the fact that the curve

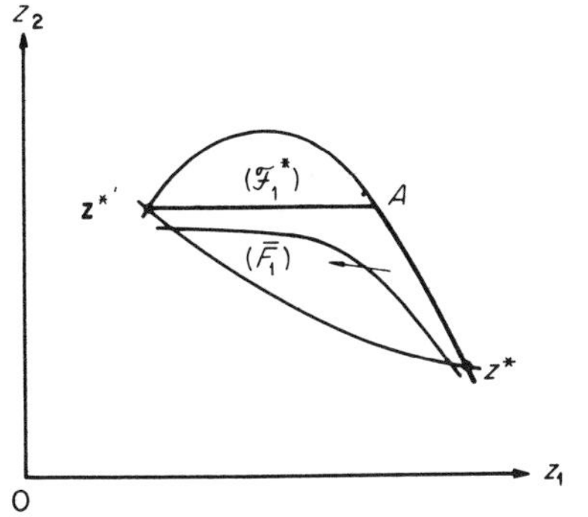

FIG. 4.2.2.d. Situation for ϵ_1/ϵ_2 small.

$(F_1{}^*)$ cannot be monotonic. In the case where the arc $(F_1{}^*)$ is monotonic, the conclusions are analogous. When this arc is not monotonic, we introduce the arc $(\mathscr{F}_1{}^*)$ defined by the monotonic function in $(z_1{}^*, z_1^{*\prime})$

$$z_2 = \mathscr{F}_1(z_1) = \inf_{z_1{}^{*\prime} \leqslant t \leqslant z_1} f_1(t), \qquad z_1^{*\prime} \leqslant z_1 \leqslant z_1{}^*$$

All the shock curves will lie between the arc $(\mathscr{F}_1{}^*)$ joining the points $\mathbf{z}^*$ and $\mathbf{z}^{*\prime}$, and the arc $(F_2{}^*)$, because on a shock curve the slope is negative. The same reasoning as before proves that, for small enough values of ϵ_1/ϵ_2, the shock curve $S(\epsilon_1, \epsilon_2)$ will be in the neighborhood of $(\mathscr{F}_1{}^*)$, given in advance.

When ϵ_2 is zero and ϵ_1 bounded, the shock layer is the arc $(F_2{}^*)$, and the value of z_1 as a function of x is given by the equation

$$\epsilon_1\, dz_1/dx = F_1(z_1, z_2) \qquad \text{where} \quad z_2 = f_2(z_1)$$

When ϵ_1 is zero and ϵ_2 is bounded, the shock layer is the arc $(\mathscr{F}_1{}^*)$ and the value of z_2 as a function of x is given by the equation

$$\epsilon_2\, dz_2/dx = F_2(z_1, z_2) \qquad \text{for} \quad z_2 < z_2^{*\prime}$$

and when $z_2 = f_1(z_1)$, then $z_1 = f_1^{-1}(z_2)$ (inverse function). The integral curve starts from the point z^* at $x = -\infty$ and moves along the arc (F_1) up to the point A of ordinate $z_2^{*\prime}$, reached at a finite value of x (which may be taken as zero); then, since z_2 is constant, dz_2 is zero and so is $F_2(z_1, z_2)$; hence, we go immediately to the point $\mathbf{z}^{*\prime}$ (see Figs. 4.2.2.e and 4.2.2.f). At $x = 0$, we have a discontinuity

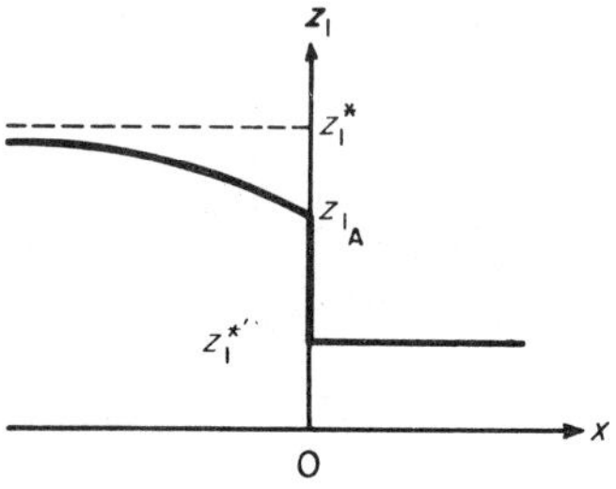

FIG. 4.2.2.e. Phenomenon of subshock.

in z_1, with z_2 remaining continuous; we call this type of discontinuity a subshock.

Now, when ϵ_2 is bounded with ϵ_1 increasing indefinitely, Eqs. (4.2.2-1) become

$$dz_1/dx = 0, \qquad \epsilon_2\, dz_2/dx = F_2(z_1, z_2)$$

z_1 stays constant. If, on the other hand, ϵ_1 stays bounded while ϵ_2 increases indefinitely, Eqs. (4.2.2-1) become

$$\epsilon_1\, dz_1/dx = F_1(z_1, z_2), \qquad dz_2/dx = 0$$

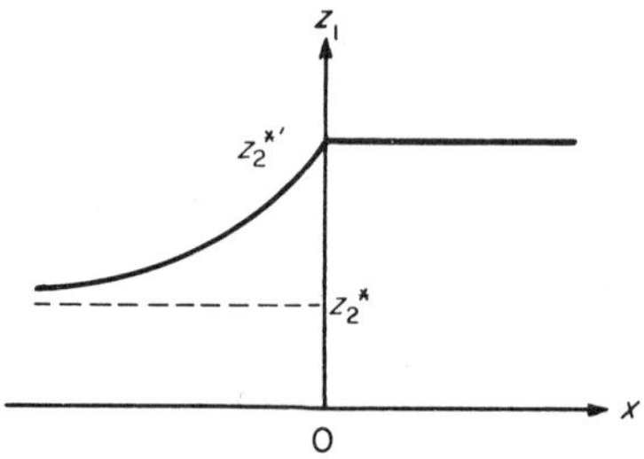

FIG. 4.2.2.f. Phenomenon of subshock.

z_2 stays constant. If the constant value is $z_2^{*\prime}$, for example, the integral curve is the segment $A\mathbf{z}^{*\prime}$; the point A corresponds to $x = -\infty$, and the point $\mathbf{z}^{*\prime}$ to $x = +\infty$.

4.2.3. Shock Structure in Nonreactive Media

It is interesting to apply the previous results to the different cases that can occur in MFD. The equations of motion are given by Eqs. (4.1.1-2), the singular points being found by putting the right-hand side equal to zero; we recover the shock equations. Assuming the Weyl inequalities to be satisfied, we will find at most four singular points. At a shock, the plane $(\mathbf{n}, \mathbf{H})$ is invariant, so that, assuming $H_z(-\infty) = 0$ (which is always possible), we will also have $H_z(+\infty) = 0$ and $w(-\infty) = w(+\infty)$, since the component of velocity perpendicular to the plane $(\mathbf{n}, \mathbf{H})$ is conserved. Naturally, $H_z(x)$ is not necessarily always zero, nor is $w(x)$ necessarily constant, but we will limit our study to the case where $H_z(x) \equiv 0$ and $w(x) = \text{const}$.

The four general equations of the shock layers will be

$$
\begin{aligned}
(\lambda_1 + 2\mu_1)\frac{du}{dx} &= mu + p + \frac{\mu H_y{}^2}{2} - P \\
\mu_1 \frac{dv}{dx} &= mv - \mu H_x H_y - P_1 \\
\lambda \frac{dT}{dx} &= m\left(\epsilon_1 - \frac{u^2 + v^2}{2}\right) - u\frac{\mu H_y{}^2}{2} + v\mu H_x H_y \\
&\quad + uP + vP_1 - E_z H_y - C \\
\sigma^{-1}\frac{dH_y}{dx} &= E_z + u\mu H_y - v\mu H_x
\end{aligned} \tag{4.2.3-1}
$$

This system has been studied by M. Germain, who proved the existence of a solution of shock structure for fast shocks $1 \to 2$ and for the slow shocks $3 \to 4$. We will limit ourselves to the two elementary cases: when the flow is rectilinear, and when one of the two coefficients λ, σ is zero. Since the motion is rectilinear, we will have $v = 0$, and consequently $H_x = 0$ or $H_y = \text{const}$.

a. First Case, $\sigma = 0$

In this case, H_y is constant, and, assuming the gas to be polytropic, the shock structure equations will be

$$(\lambda_1 + 2\mu_1)\frac{du}{dx} = mu + \frac{mRT}{u} - C_1$$
$$\lambda\frac{dT}{dx} = m\left(c_V T - \frac{u^2}{2}\right) + C_1 u - C_2 \tag{4.2.3-2}$$

where

$$C_1 = mu^* + \frac{mRT^*}{u^*}, \qquad C_2 = m\left(\frac{\gamma RT^*}{\gamma - 1} + \frac{u^{*2}}{2}\right)$$

This is the case studied in Section 4.2.2 with $u = z_1$, $T = z_2$. The hypotheses $(\mathfrak{H}_1)$-$(\mathfrak{H}_4)$ are satisfied. The system (4.2.3-2) has two singular points:

The point $u = u^*$, $T = T^*$, at which the equation for the eigenvalues of the matrix $A(\mathbf{z}^*)$ given by formula (4.2.1-2) will reduce to

$$g^2 - g\left(\mathscr{P} + \gamma - \frac{1}{M^{*2}}\right) + \gamma\mathscr{P}\left(1 - \frac{1}{M^{*2}}\right) = 0^\dagger$$

$$\mathscr{P} = \frac{\lambda_1 + 2\mu_1}{\lambda} c_p \qquad \text{(Prandtl number)}$$

$$M^{*2} = \frac{u^{*2}}{\gamma RT^*} \qquad \text{(Mach number)}$$

and the point

$$u^{*\prime} = \frac{2}{\gamma + 1}\frac{u^*}{M^{*2}} + \frac{\gamma - 1}{\gamma + 1}u^*$$

$$T^{*\prime} = T^*\frac{u^{*\prime}}{u^*}\left(\frac{2\gamma}{\gamma + 1}M^{*2} - \frac{\gamma - 1}{\gamma + 1}\right)$$

for which the equation giving the eigenvalues of $A(\mathbf{z}^{*\prime})$ becomes

$$g^2 - \left(\mathscr{P}\frac{u^{*\prime}}{u^*} + \frac{3\gamma - 1}{\gamma + 1}\frac{1}{M^{*2}} + \gamma\frac{\gamma - 3}{\gamma + 1}\right)g - \gamma\mathscr{P}\frac{u^{*\prime}}{u^*}\left(1 - \frac{1}{M^{*2}}\right) = 0$$

The Mach numbers M^* and $M^{*\prime}$ at the singular points are related by the equation

$$2\gamma(M^{*2} - 1)(M^{*\prime 2} - 1) + \gamma + 1 = 0$$

† Strictly speaking, the eigenvalues are $G = mg/\gamma(\lambda_1 + 2\mu_1)$.

so that one of them has an absolute value greater than unity. We can always assume that the singular point for which the absolute value of the Mach number is greater than unity is the one called $\mathbf{z}^*$. As $\gamma > 1$, the inequality $M^{*2} > 1$ implies that the eigenvalues $g_1{}^*$ and $g_2{}^*$ are real and positive. The eigenvalues $g_1^{*\prime}$ and $g_2^{*\prime}$ are real and of opposite sign.

Hence, the integral curve goes from $\mathbf{z}^*$ at $x = -\infty$ and reaches the point $\mathbf{z}^{*\prime}$ at $x = +\infty$, and we have $M^{*2} > 1 > M^{*\prime 2}$.

These inequalities prove that the specific entropy has a positive total variation between the points $x = -\infty$ and $x = +\infty$. For such flows, the increase of entropy (global) is no longer a hypothesis, but a property. Between positive and negative infinity, the variation is not necessarily monotonic.

When the Prandtl number is equal to one, the shock curve equation will be

$$\frac{T}{T^*} = 1 - \frac{\gamma - 1}{2} M^{*2} \left\{ \left(\frac{u}{u^*} \right)^2 - 1 \right\}$$

When the viscosity becomes zero ($\lambda_1 + 2\mu_1 = 0$), we can obtain the subshock phenomena (Fig. 4.2.3.a). For this, we require that the maximum temperature on the curve

$$u - u^* + \frac{RT}{u} - \frac{RT^*}{u^*} = 0$$

be reached between the singular points

$$2u^{*\prime} < u^* + \frac{RT}{u^*} < 2u^*$$

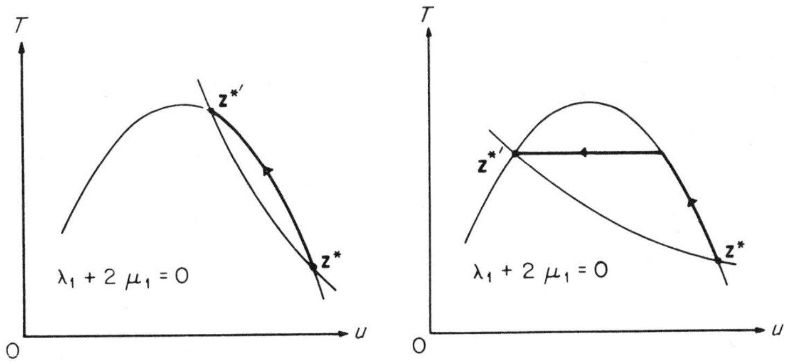

FIG. 4.2.3.a. Thermal subshock. Left: $M^{*2} \leqslant 3\gamma - 1/\gamma(3 - \gamma)$; Right: $M^{*2} > 3\gamma - 1/\gamma(3 - \gamma)$.

i.e.,

$$M^{*2} > \frac{3\gamma - 1}{\gamma(3 - \gamma)}$$

This subshock, due to the coefficient of thermal conductivity λ, is called a thermal subshock.

b. Second Case, $\lambda = 0$

When $\lambda = 0$, the temperature is a secondary variable determined by the equation

$$mc_V(T - T^*) = \tfrac{1}{2}m(u - u^*)^2 - p^*(u - u^*) + \tfrac{1}{2}u\mu(H_y - H_y{}^*)^2 + \mu H_y{}^*(H_y - H_y{}^*)(u - u^*)$$

and the shock structure equations are written

$$\begin{aligned} (\lambda_1 + 2\mu_1)\, du/dx &= P(u, H_y)/u \\ \sigma^{-1}\, dH_y/dx &= u\mu H_y - u^*\mu H_y{}^* \end{aligned} \qquad (4.2.3\text{-}3)$$

where $P(u, H_y)$ is a polynomial of the third degree,

$$P(u, H_y) \equiv \tfrac{1}{2}\mu H_y{}^2 u + mRT + mu^2 - u(mu^* + p^* + \tfrac{1}{2}\mu H_y^{*2})$$

The system (4.2.3-3) has three singular points, which are the points of intersection of the cubic $P(u, H_y)$ and the hyperbola $uH_y - u^*H_y{}^* = 0$. When the viscosity is zero, we can still have the

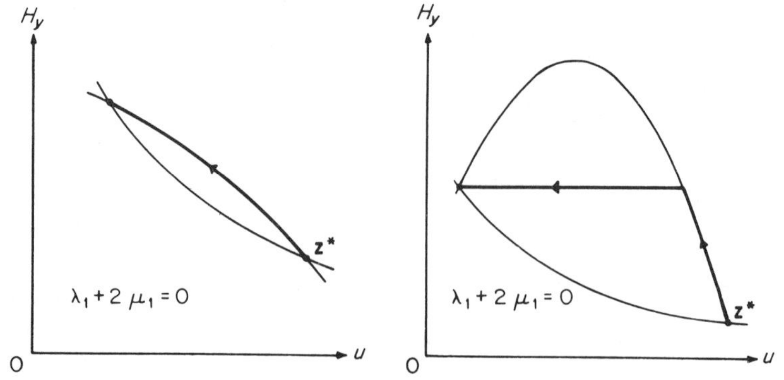

FIG. 4.2.3.b. Magnetic subshock. Left: strong magnetic field; right: weak magnetic field.

subshock phenomenon if the magnetic field $H_y{}^*$ is weak (see Fig. 4.2.3.b). On this subshock, called a magnetic subshock, H_y stays constant, while u and T have discontinuities such that

$$mu + p + \tfrac{1}{2}\mu H_y{}^2 - P \qquad \text{and} \qquad m(\epsilon_i - \tfrac{1}{2}u^2) - \tfrac{1}{2}u\mu H_y{}^2 + uP - E_z H_y - C$$

remain invariant, i.e., the magnetic subshock is a shock of classical aerodynamics.

4.2.4. Shock Structure in Reactive Media

In the case of reactive media, the transition between the uniform states far downstream and far upstream is accompanied by a change in the equation of state of the gas; we must then assume a model to describe this change. We will write

$$\epsilon_1(p, \rho) = \frac{1}{\gamma_1 - 1}\frac{p}{\rho} + C_1 \qquad \text{for} \quad x = -\infty$$

$$\epsilon_2(p, \rho) = \frac{1}{\gamma_2 - 1}\frac{p}{\rho} + C_2 \qquad \text{for} \quad x = +\infty$$

The first expression for the specific internal energy corresponds to the unburnt gas, and the second to the burnt gas. Between these two states, we write

$$\epsilon_i(p, \rho) = (1 - \alpha)\,\epsilon_1 + \alpha\,\epsilon_2$$

where α is a function of x, equal to zero for $x = -\infty$ and to unity when $x = +\infty$; $\alpha(x)$ is the proportion of burnt gas. We will assume for the variation of $\alpha(x)$ the law

$$u\, d\alpha/dx = (1 - \alpha)\, S(T) \tag{4.2.4-1}$$

which states that the speed of combustion is proportional to the quantity of unburnt gas; the coefficient of proportionality is a given function of the temperature such that

$$\begin{aligned} S(T) &= 0 && \text{for} \quad T < \bar{T} \\ &= S'(T) \geqslant 0 && \text{for} \quad T \geqslant \bar{T} \end{aligned}$$

The temperature $\bar{T}$ at which combustion starts is called the ignition temperature.

The fluid motion is determined by Eqs. (4.2.3-1) and (4.2.4-1). At infinity, the flow is uniform, so that the right-hand sides of Eqs. (4.2.3-1) are zero, and $\alpha = 0$ for $x = -\infty$, $\alpha = 1$ for $x = +\infty$. The states defined in this way correspond to the points introduced in Section 3.1-6, S_i for $x = -\infty$ and S_j' for $x = +\infty$. As the functions $\epsilon_1(p, \rho)$ and $\epsilon_2(p, \rho)$ are linear functions of temperature, there are at most four points S_i and four points S_j'. The integrals in the neighborhood of the singular points are investigated by the method of linearization explained in Section 4.2.1. The number r of the positive eigenvalues at the points S_i and the number r' of the negative eigenvalues at the points S_j are as shown in Fig. 4.2.4.a.

	S_1	S_2	S_3	S_4
r	4	3	3	2

	S_1'	S_2'	S_3'	S_4'
r'	1	2	2	3

FIG. 4.2.4.a. Number of positive (r) and negative (r') eigenvalues at S_i and S_j'.

It follows that, for the eight types of shock defined in Section 3.1.6, the degree of overdetermination of the problem of shock structure has the values $6 - (r + r')$, which gives 0 for strong detonations 1 for weak detonations and deflagrations, and 2 for strong deflagrations.

When the structure problem is overdetermined, a solution exists only if the data satisfy one additional relation in the case of weak waves, and two additional relations for strong deflagrations. The supplementary relation corresponding to weak waves can be regarded as fixing the value of $S(T)$, which cannot be arbitrary; this function can be interpreted as giving the speed of propagation of combustion, since we have

$$\alpha(t) = 1 - \exp\left\{-\int^t S(T)\, dt\right\}$$

It is, moreover, noteworthy that the degree of overdetermination of the structure problem is equal to the degree of underdetermination obtained for the piston problem given in Section 3.3.4. Hence, the missing equation for the piston problem in Section 3.3.4 when one of the shock waves is a weak detonation or deflagration can be considered as given by the structure problem: weak deflagrations and detonations propagate with a velocity which is not determined by the boundary conditions, but by the physical and chemical properties of the medium.

To clarify the preceding results, we need to examine the existence of solutions of the structure problem for all possible types of combustion waves. We will consider the simple case where $\lambda_1 = \mu_1 = \lambda = 0$, $\gamma_1 = \gamma_2 = 5/3$, $C_1 = 0$, $C_2 = -e_0$. We are then led to the solution of the following system:

$$\sigma^{-1}\frac{dH_y}{dx} = E_z + u\mu H_y$$
$$\frac{d\alpha}{dx} = \frac{1-\alpha}{u} S(T) \tag{4.2.4-2}$$

where

$$\alpha e_0 + \frac{u^2}{2} + \frac{E_z H_y}{m} + C + u\left(u^* + \frac{RT^*}{u^*} + \frac{\mu H_y^{*2}}{2m} - \frac{\mu H_y{}^2}{2m}\right) = 0 \tag{4.2.4-3}$$

and

$$RT = u\left\{\frac{RT^*}{u^*} - m(u - u^*) - \frac{\mu H_y{}^2 - \mu H_y^{*2}}{2}\right\}$$

The end points of the solutions in the space u, H_y, α are the intersections of the following surfaces: the surface (Γ) given by Eq. (4.2.4-3), the cylinder $E_z + u\mu H_y = 0$, the plane $\alpha = 0$ for $x = -\infty$ (points S_i), and the plane $\alpha = 1$ for $x = +\infty$ (points S_j'). The integral curves on (Γ) are shown in Fig. 4.2.4.b. To fix ideas, let us consider two examples

a. Fast, Strong Detonation ($S_1 \rightarrow S_2'$)

At the point S_1, one of the eigenvalues is zero and the other one is positive; at S_2', the two eigenvalues are either positive or of opposite signs. At S_1, the temperature is assumed to be smaller than the critical temperature $\bar{T}$; hence, α stays zero as long as we have $T < \bar{T}$ and the integral curve lies a long the intersection of the surface (Γ) with the plane $\alpha = 0$; we will consider the curve directed toward S_2 and along which T increases. At some point A (ahead of S_2), the

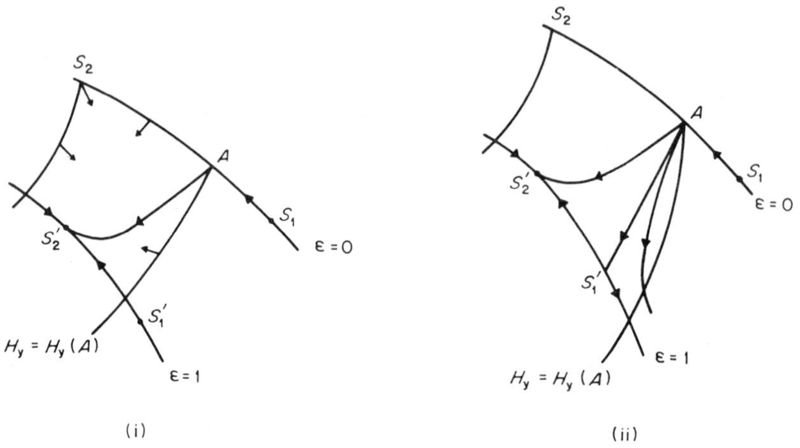

FIG. 4.2.4.b (i) Fast, strong detonation; (ii) fast, weak detonation.

ignition temperature is reached and the integral curve leaves the plane $\alpha = 0$ while remaining on the surface (Γ). We can then verify that the integral curve cannot leave the domain $(\mathscr{R})$ of the surface (Γ) bounded by the four planes $\alpha = 0$, $\alpha = 1$, $H_y = H_y(A)$, and $H_y = H_y(S_2)$. If S_2' is the only singular point on the boundary $\alpha = 1$, the integral curve reaches S_2', which proves the existence of a solution for fast, strong detonations [Fig. 4.2.4.b(i)], provided certain inequalities are satisfied by the temperature and magnetic field.

b. *Fast, Weak Detonation* $(S_1 \to S_1')$

At the point S_1', the two eigenvalues have opposite signs. If we assume that S_1' lies on the boundary $\alpha = 1$ of the region $(\mathscr{R})$, the integral curve either will still reach S_2' or will leave the region $(\mathscr{R})$ by the boundary $H_y = H_y(A)$ or, exceptionally, will reach the point S_1' [Fig. 4.2.4.b(ii)]. If this last case is to occur, the integral starting from S_1' for $x = +\infty$ must reach A when x decreases; this can only happen if a certain relation

$$f\{S, \bar{T}, E_z, H_y^*, e_0, \sigma\} = 0$$

is satisfied. We can regard this relation as determining the function $S(T)$ and say that the fast, weak detonation propagates at a speed which is independent of the boundary conditions and depends only on the physical and chemical properties of the gas.

Treating all the other cases in the same way, Soubaramayer (1967) reached the following conclusions: Strong detonations exist in general (i.e., if the ignition temperature satisfies certain inequalities); strong deflagrations do not exist; weak detonations and deflagrations exist, but propagate with fixed velocities.

4.2.5. Theory of Ionizing Shocks

A particular interesting case of shock structure in MFD occurs when the coefficient of electrical conductivity is a function of temperature $\sigma(T)$ such that σ is very small ahead of the shock and very large behind it:

$$\begin{aligned} \sigma(T) &= \sigma_1(T) \qquad \text{for} \quad T < \bar{T} \\ &= \sigma_2(T) \qquad \text{for} \quad T \geqslant \bar{T} \end{aligned}$$

This problem corresponds to the second case studied in Section 4.2.3, governed by the equations of motion (4.2.3-2), to which we must add a relation giving T as a function of u and H_y.

The temperature $\bar{T}$, called the ionizing temperature, is given and assumed to have a value between its values upstream and downstream. The shock curve is shown in Fig. 4.2.5.a; it joins the singular points $\mathbf{z}^*$ and $\mathbf{z}^{*\prime}$ of the system (4.2.3-2).

We shall look for the limiting form of this curve when the function $\sigma_1(T)$ goes to zero, while $\sigma_2(T)$ increases indefinitely. In the plane u, H_y, the equation $T = \bar{T}$ determines a certain curve which divides the plane into two regions, in one of which, $T < \bar{T}$, and in the other, $T > \bar{T}$. The limiting form of the shock curve in the second region is the arc $A'\mathbf{z}^{*\prime}$ of the hyperbola $u\mu H_y - u^*\mu H_y^* = 0$, bounded by the singular point $\mathbf{z}^{*\prime}$ and the point A', where $T = \bar{T}$. In the first region,

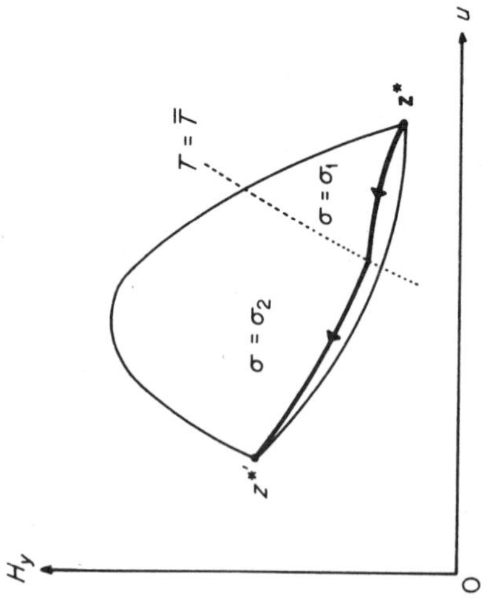

FIG. 4.2.5.a. Shock curve for electrical conductivity as a function of temperature.

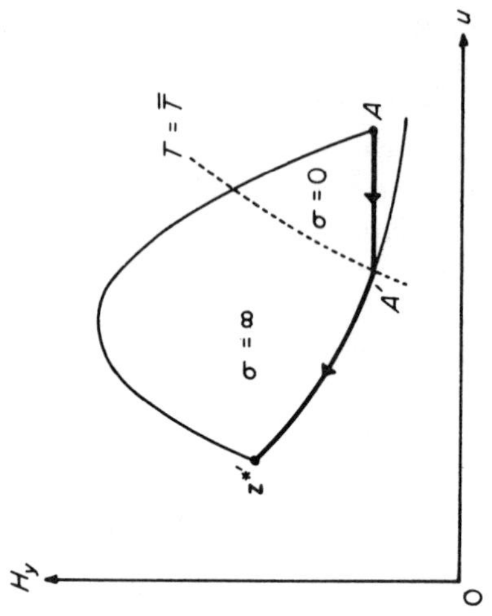

FIG. 4.2.5.b. Ionizing shock.

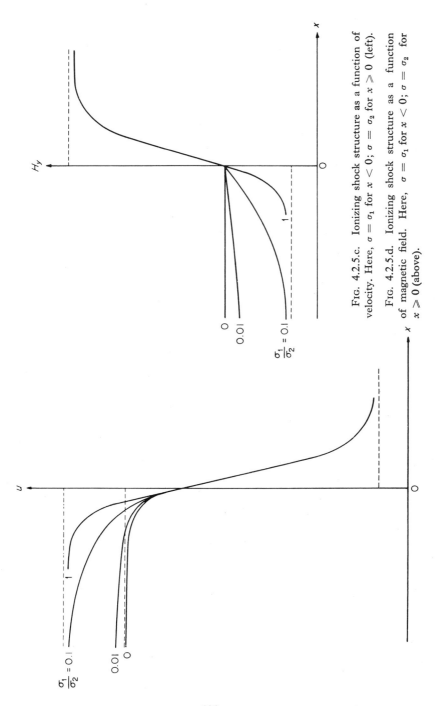

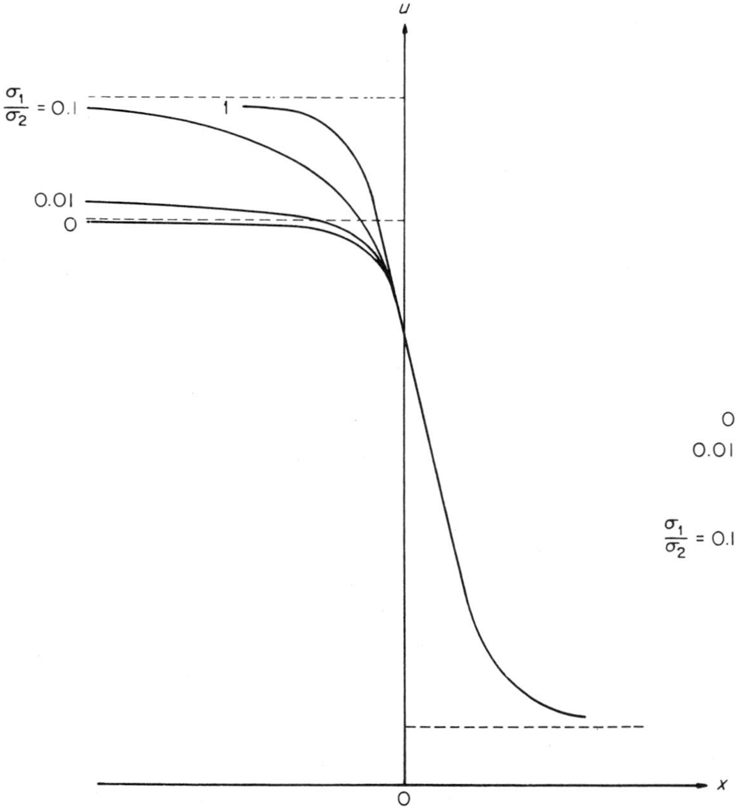

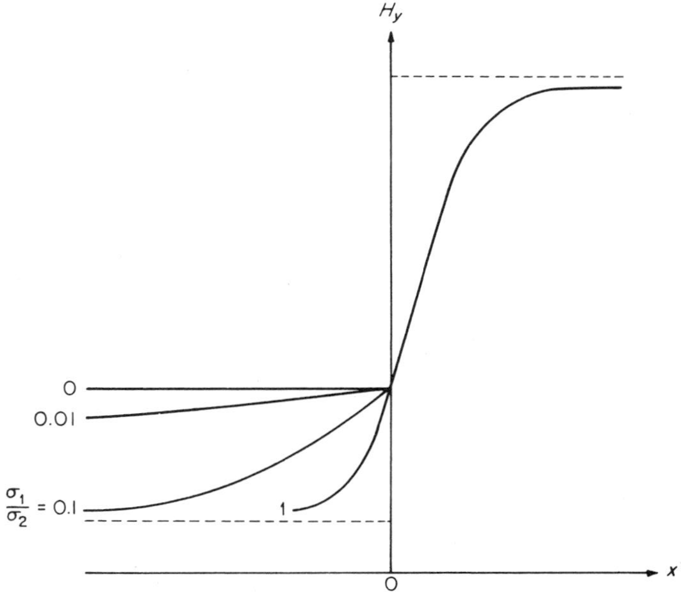

FIG. 4.2.5.c. Ionizing shock structure as a function of velocity. Here, $\sigma = \sigma_1$ for $x < 0$; $\sigma = \sigma_2$ for $x \geqslant 0$ (left).

FIG. 4.2.5.d. Ionizing shock structure as a function of magnetic field. Here, $\sigma = \sigma_1$ for $x < 0$; $\sigma = \sigma_2$ for $x \geqslant 0$ (above).

we get the case studied at the end of Section 4.2.2; the coefficient $\lambda_1 + 2\mu_1$ is bounded, while σ^{-1} increases indefinitely; on the limiting curve, H_y is constant and the point $x = -\infty$ is at A, lying on the cubic $P(u, H_y) = 0$ where $H_y = H_y(A')$ (see Fig. 4.2.5.b).

The variation of the velocity and the magnetic field as functions of x are shown for different values of σ_1/σ_2 in Fig. 4.2.5.c and 4.2.5.d.

As long as the quotient σ_1/σ_2 is not zero, the curves $u = u(x)$ have the same asymptote $u = u^*$ at $x = -\infty$; when σ_1/σ_2 becomes zero, the asymptote becomes the straight line $u = u(A)$. We find similar behavior for the curves $H_y = H_y(x)$, but in the case $\sigma_1/\sigma_2 = 0$, this curve becomes the straight line $H_y = H_y(A)$ for $x \leqslant 0$.

The relations between the coordinates of A and $\mathbf{z}^{*\prime}$ are the ionizing shock equations; these equations involve the ionizing temperature $\bar{T}$.

4.3. Nozzle Flows

4.3.1. Perfect Conductor and Zero Magnetic Field

The unsteady motion of a fluid in a nozzle of slowly varying cross section is a one-dimensional flow satisfying the equations established in Section 4.1.2. We shall consider the case of a nozzle of homothetic cross section of area $Q(x)$. When the electrical conductivity is infinite, and viscosity and thermal conductivity are negligible, the equations of motion become

$$\rho\left(\frac{\partial u}{\partial t} + u\frac{\partial u}{\partial x}\right) + \frac{\partial p}{\partial x} + \mu H_y \frac{\partial H_y}{\partial x} = 0$$

$$\frac{\partial(\rho Q)}{\partial t} + \frac{\partial(u\rho Q)}{\partial x} = 0$$

$$\frac{\partial S}{\partial t} + u\frac{\partial S}{\partial x} = 0 \qquad (4.3.1\text{-}1)$$

$$\frac{\partial H_y}{\partial t} + \frac{1}{\sqrt{Q}}\frac{\partial}{\partial x}(uH_y\sqrt{Q}) = 0$$

The third equation states that the specific entropy remains constant when following the fluid; since the specific entropy is assumed to be constant initially, it will then be constant for all x and t. Assuming the gas to be polytropic, we will have

$$p = \rho^{\gamma} \exp \frac{S - S_0}{c_V}, \qquad \frac{S - S_0}{c_V} = \text{const}$$

Equations (4.3.1-1) have solutions for which the magnetic field remains zero; these will be investigated in this section.

Let us write

$$c^2 = \frac{\partial p}{\partial \rho} = \gamma \frac{p}{\rho}, \qquad c > 0$$

$$r = u - \frac{2c}{\gamma - 1}, \qquad s = u + \frac{2c}{\gamma - 1}$$

The equations of motion then reduce to

$$\frac{D_- r}{Dt} = c\,uA, \qquad \frac{D_+ s}{Dt} = -c\,uA \tag{4.3.1-2}$$

where

$$\frac{D_-}{Dt} = \frac{\partial}{\partial t} + (u - c)\frac{\partial}{\partial x}, \quad \frac{D_+}{Dt} = \frac{\partial}{\partial t} + (u + c)\frac{\partial}{\partial x}, \quad A(x) = \frac{d[\log Q(x)]}{dx}$$

Equations (4.3.1-1) have steady solutions. In this case,

$$\rho u Q = \text{const} \qquad \text{and} \qquad u^2 + \frac{2}{\gamma - 1} c^2 = \text{const}$$

Denoting the last constant by ${u_m}^2$, we have

$$\frac{\rho}{\rho_0} = \left(1 - \frac{u^2}{{u_m}^2}\right)^{2/(\gamma-1)} \qquad \text{and} \qquad \frac{p}{p_0} = \left(1 - \frac{u^2}{{u_m}^2}\right)^{2\gamma/(\gamma-1)}$$

The flow is subsonic ($u^2 < c^2$) for $u^2 < u^{*2} = [(\gamma - 1)/(\gamma + 1)]\, {u_m}^2$ and supersonic for $u^2 > u^{*2}$. Taking the speed u to be positive, the pressure is a decreasing function of u (Fig. 4.3.1.a), while the product ρu is an increasing function for $u < u^*$ and a decreasing function for $u > u^*$ (Fig. 4.3.1.b). Knowing that $\rho u Q$ is constant, ρu will have a maximum when $Q(x)$ has a minimum, i.e., at the throat (Fig. 4.3.1.c). If the motion is continuous, subsonic in one part and supersonic in the other, sonic speed is reached at the throat (curve 3 of Fig. 4.3.1.d). If the flow is entirely subsonic, u has a maximum and p has

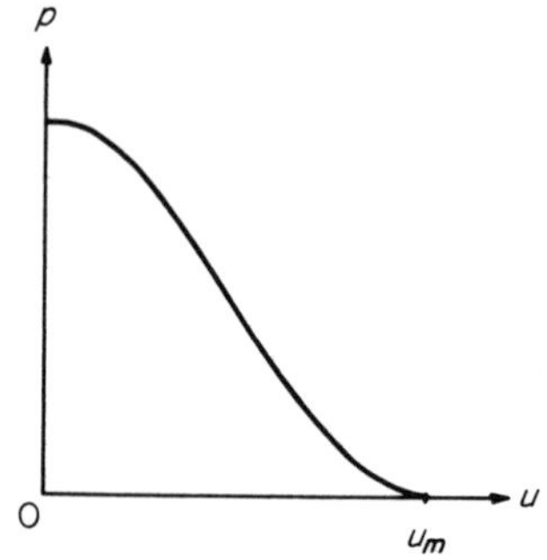

FIG. 4.3.1.a. Pressure a decreasing function of speed u (positive).

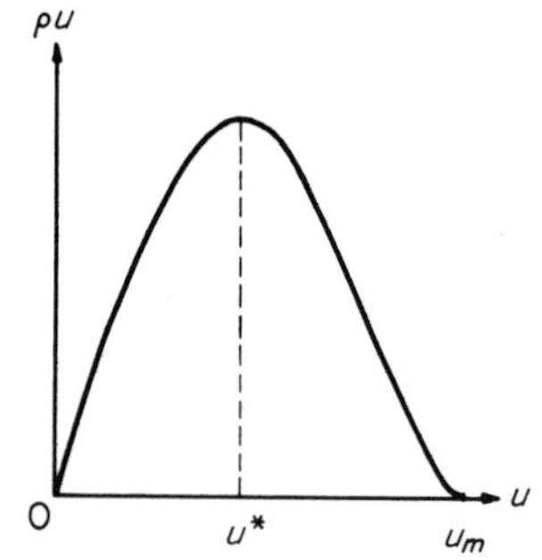

FIG. 4.3.1.b. Variation of the product ρu versus u.

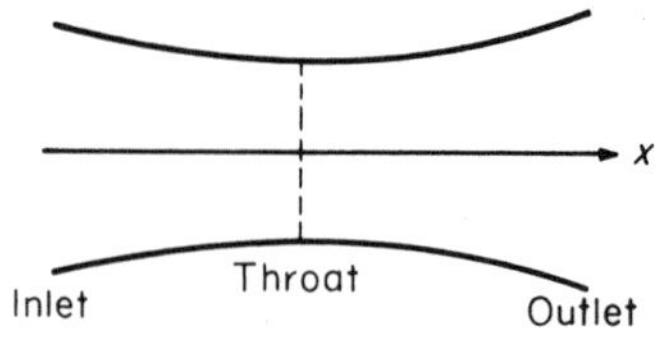

FIG. 4.3.1.c. Minimum of $Q(x)$ at throat.

a minimum at the throat (curve 1 of Fig. 4.3.1.d). If the motion is entirely supersonic, u has a minimum and p has a maximum at the throat. Finally, the flow can be subsonic ahead of the throat, supersonic beyond it, then pass through a shock which brings it back to subsonic (curve 2). In the first case, the flow is either accelerated (subsonic ahead of the throat, supersonic beyond it) or decelerated (supersonic ahead of the throat, subsonic beyond it).

A decelerated flow is unstable; in fact, any small disturbances

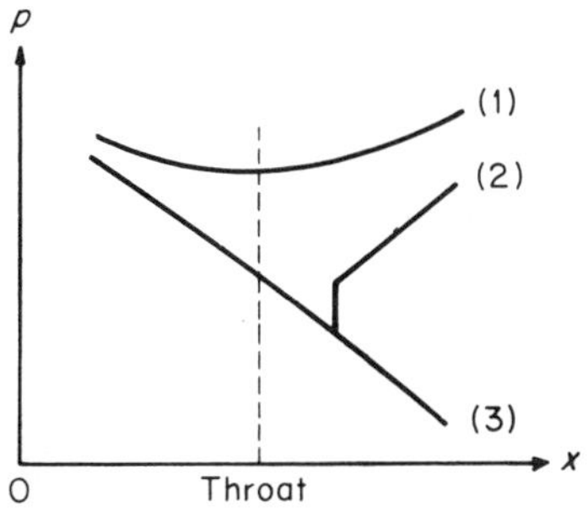

FIG. 4.3.1.d. Curve 1: u maximum, p minimum at throat; curve 2: return of flow to subsonic; curve 3: sonic speed reached at throat.

introduced in the fluid propagate along characteristics—forward waves downstream, and backward waves upstream. Forward waves introduced at the entry penetrate the nozzle and leave it after a finite time; those introduced at the exit do not enter the nozzle. All backward waves enter the nozzle, but cannot go beyond the sonic throat; they accumulate at the throat and cause a shock to be formed.

The stability of accelerated flows depends on the shape of the nozzle; we shall investigate this using Eq. (4.3.1-2). This investigation is based on the behavior of the derivatives $\sigma = D_+r/Dt$ along a backward wave, $\lambda = D_-S/Dt$ along a forward wave.

We start by considering the propagation of a small disturbance along a backward characteristic in steady flow; we use capitals for the unknowns of the steady flow: U, C, R, S, Σ,... . Differentiating the first of Eqs. (4.3.1-2), we get

$$\frac{D_+}{Dt}\left(\frac{D_-r}{Dt}\right) = A\,\frac{D_+(uc)}{Dt} + uc\,\frac{D_+A}{Dt}$$

As

$$\frac{D_+}{Dt}\left(\frac{D_-r}{Dt}\right) - \frac{D_-}{Dt}\left(\frac{D_+r}{Dt}\right) = \frac{\partial r}{\partial x}\left\{\frac{D_+(u-c)}{Dt} - \frac{D_-(u+c)}{Dt}\right\}$$

we can deduce that σ is a solution of the following Riccati-type equation, for which Σ is a particular solution:

$$\frac{D_-\sigma}{Dt} = -\,\frac{\gamma+1}{8}\,\frac{\sigma}{C}\,\{\sigma + F(x)\} + G(x)$$

where

$$2\Sigma + F(x) = \frac{4C^2}{\gamma + 1}\left(3 + \frac{1}{M}\right) M' \tag{4.3.1-3}$$

with $M = U/C$, the Mach number of the steady flow.

To integrate Eq. (4.3.1-3), we write

$$z = \frac{1}{\Sigma - \sigma}$$

$$\frac{D\_z}{Dt} = -\frac{\gamma + 1}{8C} + \frac{\gamma + 1}{8}\,\frac{2\Sigma + F}{C}\,z \tag{4.3.1-4}$$

$$\frac{z\sqrt{M}}{(M-1)^2} = \left\{\frac{z\sqrt{M}}{(M-1)^2}\right\}_{t=0} - \frac{\gamma + 1}{8}\int_{x_0}^{x} \frac{\sqrt{M}}{(M-1)^3}\,\frac{dx}{C^2}$$

The variation of the derivative $\lambda = D\_s/Dt$ along a progressive wave can be obtained in a similar way. Writing $y = 1/(\Lambda - \lambda)$, we get

$$\frac{y\sqrt{M}}{(M+1)^2} = \left\{\frac{y\sqrt{M}}{(M+1)^2}\right\}_{t=0} - \frac{\gamma + 1}{8}\int_{x_0}^{x} \frac{\sqrt{M}}{(M+1)^3}\,\frac{dx}{C^2} \tag{4.3.1-5}$$

The stability of steady flow in the nozzle is connected with the convergence of the integrals appearing in the right of Eqs. (4.3.1-4) and (4.3.1-5), because the value of z (or y) is related to the derivative of the pressure. We have, in fact,

$$\frac{\partial u}{\partial t} + (u + c)\frac{\partial u}{\partial x} - \frac{2}{\gamma - 1}\left\{\frac{\partial c}{\partial t} + (u + c)\frac{\partial c}{\partial x}\right\} = \sigma$$

$$\frac{\partial u}{\partial t} + (u + c)\frac{\partial u}{\partial x} + \frac{2}{\gamma - 1}\left\{\frac{\partial c}{\partial t} + (u + c)\frac{\partial c}{\partial x}\right\} = -cuA$$

and therefore

$$\sigma = -\,cuA - \frac{4}{\gamma - 1}\left\{\frac{\partial c}{\partial t} + (u + c)\frac{\partial c}{\partial x}\right\}$$

$$\Sigma = -\,cuA - \frac{4}{\gamma - 1}(u + c)\,C'$$

$$\sigma - \Sigma = -\frac{4}{\gamma - 1}\left\{\frac{\partial c}{\partial t} + (u + c)\left(\frac{\partial c}{\partial x} - C'\right)\right\}$$

If we move along a backward wave, the derivatives along this wave are continuous; in particular,

$$\frac{\partial c}{\partial t} + (u - c)\frac{\partial c}{\partial x} = (u - c)\, C'$$

Thus,

$$\sigma - \Sigma = \frac{8}{\gamma - 1}\frac{C}{U - C}\frac{\partial c}{\partial t}$$
$$\frac{\partial p}{\partial t} = -\frac{\rho C}{4}\frac{M - 1}{z} \tag{4.3.1-6}$$

Similarly, along a forward wave,

$$\frac{\partial p}{\partial t} = \frac{\rho C}{4}\frac{M + 1}{y} \tag{4.3.1-7}$$

We consider an accelerated flow. An increase of pressure in the convergent part ($M < 1$) corresponds to z_0 positive; a backward wave propagates in the direction of upstream infinity; the integral in the Eq. (4.3.1-4) is then positive, and, if it diverges, z will become zero and the derivative of the pressure becomes infinite; the flow will be unstable; if the integral converges, we can always choose z_0 large enough, i.e., a disturbance small enough, so that z is never zero; the flow will therefore be stable. Thus, for $x \to -\infty$, we have

$$U \to 0, \qquad C \to C_0, \qquad M \to 0$$

The integral converges (stable flow) if

$$\sqrt{M} \sim \frac{1}{x^{1+\epsilon}}, \qquad \epsilon > 0$$

$$x^2 M \to 0, \qquad \text{or} \qquad x^{-2} Q(x) \to \infty$$

In order to have a stable flow in a convergent nozzle, the cross-sectional area must vary more rapidly than that of a cone.

In the divergent part of the nozzle ($M > 1$), there will be stability if the integral (4.3.1-4) or (4.3.1-5) converges, but this leads to an analogous result. In fact, for $x \to \infty$, we have

$$U \to U_m, \qquad C \to 0, \qquad M \to \infty$$

The integrals will converge (stable flow) if

$$\frac{\sqrt{M}}{M^3}\frac{1}{C^2} \sim \frac{1}{x^{1+\epsilon}}, \qquad \epsilon > 0$$

$$x^2 C \to 0, \qquad \text{or} \qquad x^{-2} Q(x)^{(\gamma-1)/2} \to \infty$$

for C goes to zero like $\rho^{(\gamma-1)/2}$, or like $Q^{-(\gamma-1)/2}$, since $\rho u Q$ is constant.

In conclusion, the accelerated flow in a nozzle will be stable if the area of the normal section satisfies the equations

$$\lim_{x\to-\infty} x^2/Q(x) = 0, \qquad \text{upstream infinity}$$

$$\lim_{x\to\infty} x^2/Q(x)^{(\gamma-1)/2} = 0, \qquad \text{downstream infinity}$$

4.3.2. Perfect Conductor, Magnetic Field Nonzero

The extension of the above investigation to MFD demonstrates the influence of a magnetic field on the stability of certain flows. We find a large number of possible cases, depending on whether the nozzle is two dimensional or has homothetic sections, or whether certain walls of the nozzle are conducting or insulated.

We will study the case of a two-dimensional nozzle in a magnetic field in which the curved surfaces are insulated while the plane surfaces only are conducting and at different electrical potentials (Fig. 4.3.2.a). The plane walls are parallel to the x, y plane, the x axis being the nozzle axis.

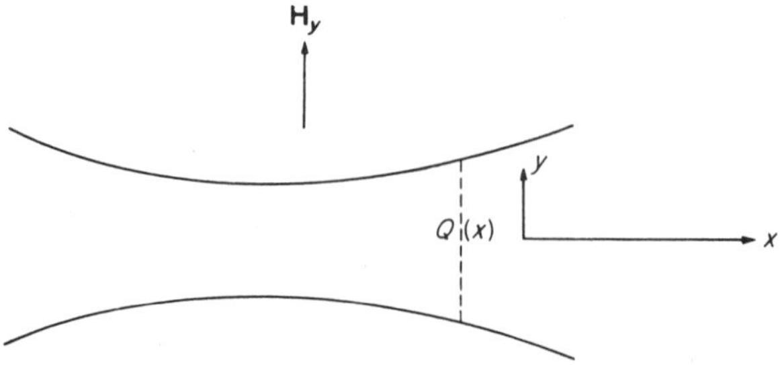

Fig. 4.3.2.a. Insulated curved walls — plane conducting walls.

The equations of motion are then

$$
\begin{aligned}
\rho\left(\frac{\partial u}{\partial t} + u\,\frac{\partial u}{\partial x}\right) + \frac{\partial p}{\partial x} + \mu H_y J_z &= 0 \\
\frac{\partial(\rho Q)}{\partial t} + \frac{\partial(u\rho Q)}{\partial x} &= 0 \\
\frac{\partial S}{\partial t} + u\,\frac{\partial S}{\partial x} &= 0 \\
\frac{\partial H_y}{\partial t} + \frac{\partial(uH_y)}{\partial x} &= 0
\end{aligned}
\tag{4.3.2-1}
$$

Combining the second and the fourth equations, we get

$$\frac{\partial}{\partial t}\left(\frac{H_y}{\rho Q}\right) + u\,\frac{\partial}{\partial x}\left(\frac{H_y}{\rho Q}\right) = 0$$

which means that the ratio $H_y/\rho Q$, like the entropy S, stays constant as we follow the fluid. If the flow is steady, these two quantities are constant; this will also be the case if the flow was steady at some earlier time, as we suppose. Hence, we can write

$$S = \text{const} \qquad \text{and} \qquad H_y/\rho Q = \text{const}$$

instead of the last two equations of the system (4.3.2-1).

From the equation $\mathbf{J} = \nabla \times \mathbf{H}$, we have $J_z = \partial H_y/\partial x - \partial H_x/\partial y$; but, since the height of the nozzle is small compared to its length, we can write

$$\frac{\partial H_x}{\partial y} = \frac{[H_x \quad \text{on wall}]}{Q(x)} = \xi(x)$$

where $\xi(x)$ can be assumed known, if the magnetic field on the wall is produced by an external process sufficiently powerful to be uninfluenced by the flow. The problem is then reduced to that of finding two functions u and ρ satisfying the equations

$$
\begin{aligned}
\frac{\partial u}{\partial t} + u\,\frac{\partial u}{\partial x} + \frac{1}{\rho}\,\frac{\partial p}{\partial x} &= \frac{\mu H_y}{\rho}\left(\xi - \frac{\partial H_y}{\partial x}\right) \\
\frac{\partial(\log\rho)}{\partial t} + u\,\frac{\partial(\log\rho)}{\partial x} + \frac{\partial u}{\partial x} &= -u\,\frac{d(\log Q)}{dx}
\end{aligned}
\tag{4.3.2-2}
$$

We will again assume a polytropic gas and write

$$a^2 = \frac{\partial p}{\partial \rho} = \gamma \frac{p}{\rho}, \qquad \alpha^2 = \frac{\mu H_y^2}{\rho}, \qquad c^2 = a^2 + \alpha^2; \qquad A(x) = \frac{d(\log Q)}{dx}$$

where c is the fast magnetosonic speed; small disturbances propagate with the speed $u + c$ for forward waves and $u - c$ for the backward waves. We can rewrite Eqs. (4.3.2-2) in the form

$$\begin{aligned} c\frac{D_-(\log \rho)}{Dt} - \frac{D_- u}{Dt} &= -(cu + \alpha^2)A - \frac{\mu H_y}{\rho}\xi \\ c\frac{D_+(\log \rho)}{Dt} + \frac{D_+ u}{Dt} &= -(cu + \alpha^2)A + \frac{\mu H_y}{\rho}\xi \end{aligned} \qquad (4.3.2\text{-}3)$$

These equations have a steady solution which will be denoted by capitals. Writing U, R, C as functions of x corresponding to u, ρ, c, we have

$$UU' + C^2\frac{R'}{R} = \frac{\mu H_y}{R}\xi - \alpha^2 A$$

$$RU' + UR' = -URA$$

We can deduce, in particular,

$$\frac{U'}{U}(M^2 - 1) = (1 - \beta^2)\frac{\xi}{H_y} + \beta^2 A$$

where

$$M = U/C, \qquad \beta^{-2} = (c/a)^2 = 1 + (\alpha/a)^2 > 1$$

The coefficient $1 - \beta^2$ being positive, the magnetosonic point (defined by $M = 1$) is not at the throat of the nozzle, but in the divergent part ($A > 0$) if $\xi/H_y < 0$, and in the convergent part ($A < 0$) if $\xi/H_y > 0$. The study of stability is similar to that given in the previous section. We shall consider the case of retarded steady flows and prove that flows which are unstable when the magnetic field is absent can be made stable by the introduction of a magnetic field.

The variation of the quantity $\sigma = \frac{1}{2} D_+u/Dt$ along a backward wave is determined by a differential equation of Riccati type, a particular solution of which is known; the value Σ corresponding to steady flow. The function $z = 1/(\Sigma - \sigma)$ satisfies the linear equation

$$\frac{D_- z}{Dt} = -\frac{1}{2C}\{J_1(x) + zJ_2(x)\} \qquad (4.3.2\text{-}4)$$

where

$$J_1(x) = 3 + (\gamma - 2)\,\beta^2 > 0$$

$$J_2(x) = -C^2M'(3 + 1/M) + C^2A(\beta^2 - 1)$$

The solution of Eq. (4.3.2-4) can be written in the form

$$Lz = \{Lz\}_{t=0} - \frac{1}{2}\int_{x_0}^{x} \frac{LJ_1}{C^2}\frac{dx}{M-1}$$

$$L = \exp\left\{\frac{1}{2}\int_{x_0}^{x} \frac{J_2}{C^2}\frac{dx}{M-1}\right\} \qquad (4.3.2\text{-}5)$$

Since the variation of the pressure along a backward wave is given by

$$\partial p/\partial t = -\rho C\beta^2(M-1)/z$$

the flow will be unstable if z can vanish and stable in the opposite case. When the magnetic field is zero ($\beta^2 = 1$), the function L can be given explicitly and the discussion is elementary; when the magnetic field is not zero ($\beta^2 < 1$), we must confine ourselves to the study of stability by examining the sign of $J_2(x)$ in the neighborhood of the magnetosonic point (i.e., when time increases indefinitely).

a. First Case: $J_2(x) > 0$ at the Magnetosonic Point

This is the only case that appears when the magnetic field is zero ($\beta^2 = 1$), and the flow is decelerated $M' < 0$. We can then expect similar conclusions.

An increase of pressure at the exit corresponds to $z_0 > 0$ (because $M < 1$); when x decreases from the value at the exit corresponding to the magnetosonic point, the integral

$$\frac{1}{2}\int_{x_0}^{x} \frac{LJ_1}{C^2}\frac{dx}{M-1}$$

tends to $+\infty$; hence, z becomes zero after a certain time and the flow is unstable. The fact that $J_2(x)$ is positive causes the function L to increase indefinitely as we approach the magnetosonic point.

b. Second Case: $J_2(x) < 0$ at the Magnetosonic Point

This case is interesting because it does not appear when the magnetic field is zero in a decelerated flow.

As we approach the magnetosonic point, we find

$$J_2/2C \to -k$$

where k is a positive constant. We have

$$L = \exp\left\{\int_0^t \frac{J_2}{2C}\, dt\right\} \sim \exp\{-kt\}$$

$$\Delta = \frac{1}{2}\int_0^t \frac{LJ_1}{C}\, dt \to \Delta^*$$

It is always possible to choose the initial disturbance in pressure so that $L_0 z_0$ is larger than Δ^*; thus, z does not vanish, which proves the stability of the flow.

The stability of a decelerated flow is then related to the sign of the function $J_2(z)$; the condition for neutral stability, $J_2(x) = 0$, will be

$$-M'\left(3 + \frac{1}{M}\right) + \frac{Q'}{Q}(\beta^2 - 1) = 0$$

For $\gamma = 2$, this condition can be integrated to give

$$(1 - \beta_0^2)\left(\frac{Q}{Q_0}\right)^2 = \frac{\exp\{-6(M-1)\}}{M^2} - \beta_0^2$$

The curves $M = M(Q/Q_0)$ corresponding to different values of β_0 are shown in Fig. 4.3.2.b. For a given nozzle, β_0 and Q_0 (values corresponding to the magnetosonic point) are known, as is the function $M(Q)$; if the curve representing this function lies between the straight line $\beta_0^2 = 1$ and the curve corresponding to the nozzle, the flow will be stable; this will be a decelerated flow stabilized by a magnetic field.

4.3.3. Theory of the Electromagnetic Flow Meter

As an example of flow in a straight line, let us look at the flow of an incompressible viscous fluid in a straight duct; the thermal conductivity is neglected and the electrical condutivity is arbitrary. The equations of motion have steady solutions for which the vectors

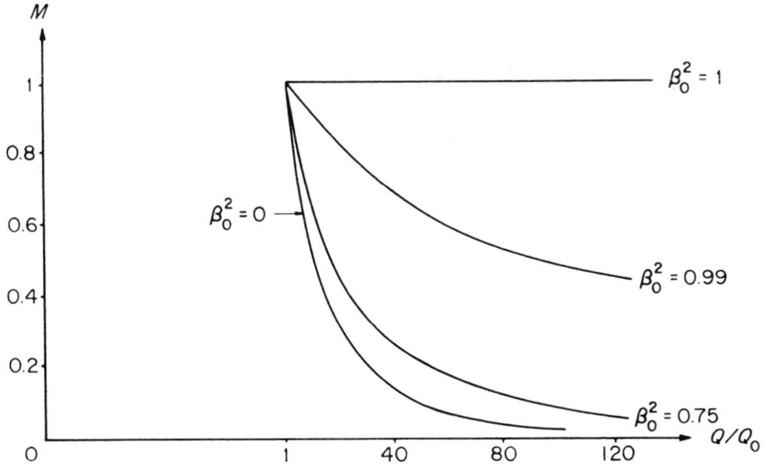

FIG. 4.3.2.b. Stability of decelerated flows in a nozzle, β_0 and Q_0 are the values of β and Q for $M = 1$.

V, **E**, and **H** depend only on the coordinates y and z and have the following components:

$$\mathbf{V} = \begin{vmatrix} u \\ 0 \\ 0 \end{vmatrix}, \qquad \mathbf{E} = \begin{vmatrix} 0 \\ E_y \\ E_z \end{vmatrix}, \qquad \mathbf{H} = \begin{vmatrix} \mathrm{h} \\ H_0\,(\mathrm{const}) \\ 0 \end{vmatrix}$$

The density S is a constant and the pressure p depends on, x y, and z. The equations of motion have the form

$$\begin{gathered} -\frac{\partial p}{\partial x} + \mu_1\left(\frac{\partial^2 u}{\partial y^2} + \frac{\partial^2 u}{\partial z^2}\right) + \mu H_0 \frac{\partial h}{\partial y} = 0 \\ \frac{\partial}{\partial y}\left(p + \frac{\mu h^2}{2}\right) = \frac{\partial}{\partial z}\left(p + \frac{\mu h^2}{2}\right) = 0 \\ \frac{\partial h}{\partial y} = -\sigma(E_z + u\mu H_0), \qquad \frac{\partial h}{\partial z} = \sigma E_y \\ \frac{\partial E_y}{\partial z} - \frac{\partial E_z}{\partial y} = 0 \end{gathered} \tag{4.3.3-1}$$

There is a first integral $P + \frac{1}{2}\mu h^2 = f(x)$, where $f(x)$ is a linear function in virtue of the first equation of (4.3.3-1). The total pressure varies linearly with x.

We denote by R some characteristic length and by V_0 some characteristic velocity and write $H_1 = V_0(\sigma\mu_1)^{1/2}$. We now introduce the dimensionless variables $Y = y/R$ and $Z = z/R$ and the dimensionless functions $v = u/V_0$ and $b = h/H_1$. The equations of motion reduce to the following:

$$\frac{\partial^2 b}{\partial Y^2} + \frac{\partial^2 b}{\partial Z^2} + \sigma\mu\frac{RV_0H_0}{H_1}\frac{\partial v}{\partial Y} = 0$$
$$\frac{\partial^2 v}{\partial Y^2} + \frac{\partial^2 v}{\partial Z^2} + \mu\frac{H_0H_1R}{V_0\mu_1}\frac{\partial b}{\partial Y} - \frac{R^2}{V_0\mu_1}\frac{\partial p}{\partial x} = 0 \qquad (4.3.3\text{-}2)$$

the terms of the first order have a common coefficient, $H = H_0R(\sigma/\mu_1)^{1/2}$, the Hartman number. Writing

$$P = -\frac{R^2}{\mu_1 V_0}\frac{\partial p}{\partial x} = \text{const}$$

and $\alpha = v + b$, $\beta = v - b$, we then have to solve the system

$$\Delta\alpha + H\frac{\partial\alpha}{\partial Y} + P = 0, \qquad \Delta\beta - H\frac{\partial\beta}{\partial Y} + P = 0 \qquad (4.3.3\text{-}3)$$

where $\Delta = \partial^2/\partial Y^2 + \partial^2/\partial Z^2$. In practice, the liquid flows in a channel with electrically insulated walls. On the boundary, we must have $v = 0$ and $J_n = 0$, which leads to $h = \text{const}$, a constant which can be taken as zero. We then have to find functions α and β which are zero on the boundary and which satisfy Eqs. (4.3.3-3).

We write

$$\alpha = -(P/H)Y + \alpha_1 e^{-HY/2}$$

Changing to cylindrical coordinates ($X = r\sin\theta$, $y = r\cos\theta$), we get

$$\frac{\partial^2\alpha_1}{\partial r^2} + \frac{1}{r}\frac{\partial\alpha_1}{\partial r} + \frac{1}{r^2}\frac{\partial^2\alpha_1}{\partial\theta^2} - \frac{H^2\alpha_1}{4} = 0$$

which allows us to treat channels of circular cross section.

If we look for the solutions in the form of Fourier series

$$\alpha_1 = \sum_{n=0}^{\infty} f_n(r)\cos n\theta$$

we get the differential equations

$$\frac{d^2 f_n}{dr^2} + \frac{1}{r}\frac{df_n}{dr} - \left(\frac{H^2}{4} + \frac{n^2}{r^2}\right) f_n = 0$$

the regular solutions of which, in the neighborhood of the origin, are Bessel functions of imaginary argument

$$f_n = a_n \frac{I_n(Hr/2)}{I_n(H/2)}$$

The constants a_n are determined by the condition on the contour $r = 1$, which can be written

$$\alpha_1(1, \theta) = \sum a_n \cos n\theta \equiv (P/H) \cos \theta \, e^{(H \cos \theta)/2}$$

Using the relation $I_n(z) = (1/2\pi) \int_0^{2\pi} e^{z\cos\theta} \cos n\theta \, d\theta$, we deduce that

$$a_0 = I_1(H/2), \qquad a_n = I_{n-1}(H/2) + I_{n+1}(H/2)$$

which determines our solution.

The constant P is related to the Hartman number (Fig. 4.3.3.a) and is calculated from the conditions that the flux is $\pi R^2 V_0$ (defining V_0 as the average velocity). We get

$$\int_0^1 r \, dr \int_0^{2\pi} v \, d\theta = \pi$$

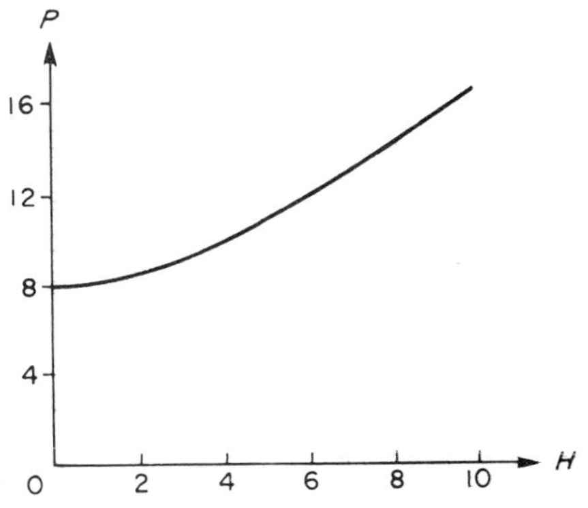

FIG. 4.3.3.a. The constant P as a function of H, the Hartman number.

whence

$$\frac{H}{P} = \frac{I_1(H/2)}{I_0(H/2)} \left\{ I_0^2\left(\frac{H}{2}\right) - I_1^2\left(\frac{H}{2}\right) \right\} + \sum_{n=1}^{\infty} (-)^n \frac{I_{n+1}(H/2) + I_{n-1}(H/2)}{I_n(H/2)}$$
$$\times \left\{ \left(1 + \frac{4n^2}{H^2}\right) I_n^2\left(\frac{H}{2}\right) - \left[\frac{I_{n+1}(H/2) + I_{n-1}(H/2)}{2}\right]^2 \right\}$$

Figure 4.3.3.b represents curves of equal velocity and Fig. 4.3.3.c represents curves of constant b/H.

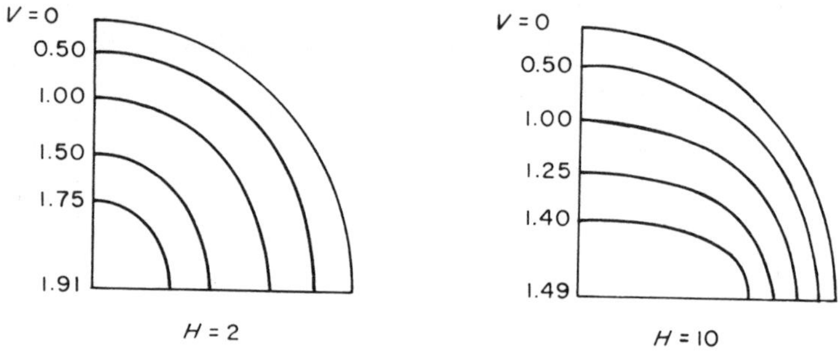

FIG. 4.3.3.b. Constant-velocity lines.

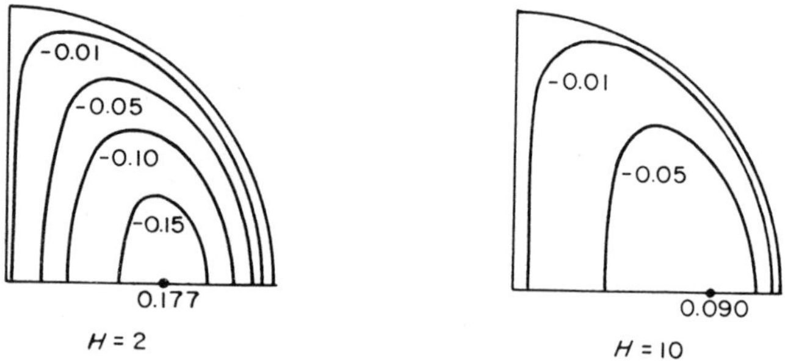

FIG. 4.3.3.c. Constant-induction lines.

CHAPTER 5

FLOW PAST BODIES

The study of the motion of a solid body in an electrically conducting fluid, subject to a magnetic field, is of great interest to workers doing research with rockets and propellers. It is hoped that, in the phenomena of magnetofluiddynamics, new possibilities can be found for modifying the load distributions and increasing thrust. The methods used to study such motions are very close to those of classical aerodynamics and, in certain cases, the equations of magnetofluiddynamics actually reduce to the aerodynamic equations; three cases of this type are studied in the first section. The following two sections are devoted to the problem of flow past a semiinfinite or finite flat plate. The first case corresponds to the case of a boundary layer which, in the case of magnetofluiddynamics, can have two solutions; in the second case, at constant Reynolds number, the drag is an increasing or decreasing function of the electrical conductivity, depending on the strength of the magnetic field. In the last two sections, the flow around a sphere is investigated; in Section 5.4, we consider the case of a fully magnetized sphere in an inviscid fluid, and in Section 5.5, the case of viscous flow around a sphere with a cavity.

5.1. Reduction of the Equations in Special Cases

5.1.1. Flows with an Aligned Magnetic Field

The equations of magnetofluiddynamics for fluids which are perfect conductors of electricity, in the absence of viscosity and heat conduction, possess steady isentropic solutions for which the magnetic field is parallel to the velocity. This type of flow is called flow with an aligned magnetic field. We write

$$\mathbf{H} = \lambda \rho \mathbf{V}$$

Then

$$\nabla \cdot \mathbf{H} = \lambda \nabla \cdot \rho \mathbf{V} + \nabla \lambda \cdot \rho \mathbf{V}$$

$$\nabla \times \mathbf{H} = \lambda \rho \nabla \times \mathbf{V} + \nabla(\lambda \rho) \times \mathbf{V}$$

and we will verify that the equations of motion

$$\rho \{(\nabla \times \mathbf{V}) \times \mathbf{V} + \nabla(V^2/2)\} = -\nabla p + (\nabla \times \mathbf{H}) \times \mu \mathbf{H}$$

$$\nabla \cdot \rho \mathbf{V} = 0, \qquad \nabla \cdot \mathbf{H} = 0, \qquad \nabla \times (\mathbf{V} \times \mathbf{H}) = 0, \qquad p = f(\rho)$$

are compatible with the hypothesis $\mathbf{H} = \lambda \rho \mathbf{V}$.

The first three equations give

$$\mathbf{V} \cdot \nabla \left\{ \frac{V^2}{2} + \int \frac{dp}{\rho} \right\} = 0, \qquad \mathbf{V} \cdot \nabla \lambda = 0$$

The quantities $V^2 + 2 \int dp/\rho$ and λ are then constant on each streamline and, assuming the flow to be uniform at infinity, where all streamlines must be parallel, we conclude the existence of closed streamlines; they will be constant throughout the flow. We then have

$$V^2 + 2 \int dp/\rho = \text{const}, \qquad \lambda = \text{const}$$

Using these two relations, the equations of motion reduce to

$$\rho \mathbf{V} \times (\nabla \times \mathbf{V}) - \mu \mathbf{H} \times (\nabla \times \mathbf{H}) = 0, \qquad \nabla \cdot \rho \mathbf{V} = 0 \quad (5.1.1\text{-}1)$$

The first equation can be written in the form

$$\rho \mathbf{V} \times \nabla \times (\mathbf{V} - \lambda \mu \mathbf{H}) = 0$$

$$\nabla \times (\mathbf{V} - \lambda \mu \mathbf{H}) = k \rho \mathbf{V}$$

Hence.

$$\nabla \cdot k \rho \mathbf{V} \equiv k \nabla \cdot \rho \mathbf{V} + \rho \mathbf{V} \cdot \nabla k = \rho \mathbf{V} \cdot \nabla k = 0$$

The function k is then constant on each streamline; but, since k is zero at infinity (because $\mathbf{H}$ and $\mathbf{V}$ are uniform), it will have to be zero throughout the flow. The system then reduces to

$$\nabla \times \{(1 - \lambda^2 \mu \rho) \mathbf{V}\} = 0, \qquad \nabla \cdot \rho \mathbf{V} = 0 \qquad (5.1.1\text{-}2)$$

The preceding equations can be interpreted as those governing a fictitious irrotational flow of a nonconducting compressible fluid.

For this reason, we introduce the Alfvén number A defined by the relation

$$A^2 = \frac{V^2}{\mu H^2/\rho} = \frac{1}{\lambda^2 \mu \rho} = A_\infty{}^2 \frac{\rho_\infty}{\rho}$$

and we write

$$\mathbf{V} = V_\infty \frac{1 - A_\infty^{-2}}{1 - A^{-2}} \mathbf{b} \tag{5.1.1-3}$$

or

$$\mathbf{H} = H_\infty \frac{1 - A_\infty{}^2}{1 - A^2} \mathbf{b}$$

Equations (5.1.1.2) become

$$\nabla \times \mathbf{b} = 0, \qquad \nabla \cdot \sigma\mathbf{b} = 0 \tag{5.1.1-4}$$

with

$$\sigma = (1 - A_\infty{}^2)/(1 - A^2) = f(b)$$

The coefficient σ is a function of $b = |\, \mathbf{b} \,|$, since we have

$$V^2 = f_1(\rho) = f_2(A^2) \qquad \text{and} \qquad b = V f_3(A^2) = f_4(A^2)$$

Equations (5.1.1-4) allow us to interpret $\mathbf{b}$ as the velocity vector of an irrotational flow of a compressible fluid with density σ varying as a function of b, $\sigma = f(b)$. The equations of magnetofluiddynamics are then reduced to the equations of aerodynamics for some fictitious fluid.

In aerodynamics, we have $dp + \rho V\, dV = 0$ and the speed of propagation of small disturbances is given by

$$c^2 = dp/d\rho = -\rho V\, dV/d\rho$$

For our imaginary gas, the speed of propagation of small disturbances, a, which we will call a pseudospeed of sound, will be given by

$$a^2 = -\sigma b\, db/d\sigma = -\tfrac{1}{2}\, db^2/(d \log \sigma)$$

The pseudospeed of sound a is proportional to b, the ratio of the two depending only on the Alfvén number A and the number $M = V/c$, which we will call the Mach number. We have, in fact,

$$b^2 = \left\{\frac{1 - A^{-2}}{1 - A_\infty^{-2}}\right\}^2 \frac{V^2}{V_\infty^{\ 2}}, \qquad \sigma = \frac{1 - A_\infty^{\ 2}}{1 - A^2}$$

$$\frac{db^2}{b^2} = 2\frac{A^{-4}\, dA^2}{1 - A^{-2}} + \frac{dV^2}{V^2}, \qquad \frac{d\sigma}{\sigma} = \frac{dA^2}{1 - A^2}$$

Thus

$$dV^2 = -2c^2 \frac{d\rho}{\rho} \quad \text{and} \quad \rho = \rho_\infty \frac{A_\infty^{\ 2}}{A^2}$$

Then

$$db^2 = 2b^2 \left\{\frac{1}{A^2M^2} + \frac{1}{A^2(A^2 - 1)}\right\} dA^2$$

and

$$a^2 = \frac{A^2 + M^2 - 1}{A^2M^2}\, b^2 \tag{5.1.1-5}$$

The first of Eqs. (5.1.1-4) admits a solution of the form $\mathbf{b} = \nabla\Phi$. The second equation will then be written

$$\nabla \cdot (\sigma\nabla\Phi) = 0, \qquad \sigma = f(b)$$

or

$$\sigma\nabla \cdot (\nabla\Phi) + \frac{d\sigma}{db} \nabla\Phi \cdot \nabla b = 0$$

Since $a^2\, d\sigma = -\sigma b\, db$, the function $\Phi(x, y, z)$ satisfies the equation

$$\sum (a^2 - b_x^{\ 2}) \frac{\partial^2\Phi}{\partial x^2} - 2 \sum b_y b_z \frac{\partial^2\Phi}{\partial y\, \partial z} = 0 \tag{5.1.1-6}$$

In the case of small disturbances, $\mathbf{b} \approx b\mathbf{x}$, the equation for the pseudovelocity potential Φ can be linearized in the form

$$(1 - m^2) \frac{\partial^2\Phi}{\partial x^2} + \frac{\partial^2\Phi}{\partial y^2} + \frac{\partial^2\Phi}{\partial z^2} = 0$$

$$(1 - m^2) = 1 - \frac{b^2}{a^2} = \frac{(1 - M^2)(A^2 - 1)}{A^2 + M^2 - 1}$$

To discuss the properties of the flow, it is convenient to use M^2 and A^2 as coordinates (see Fig. 5.1.1-a) and consider the five following regions:

(1) $A^2 > 1, \quad M^2 < 1$

(2) $A^2 < 1, \quad M^2 > 1$

(3) $A^2 > 1, \quad M^2 > 1$

(4) $A^2 + M^2 > 1, \quad A^2 < 1, \quad M^2 < 1$

(5) $A^2 + M^2 < 1$

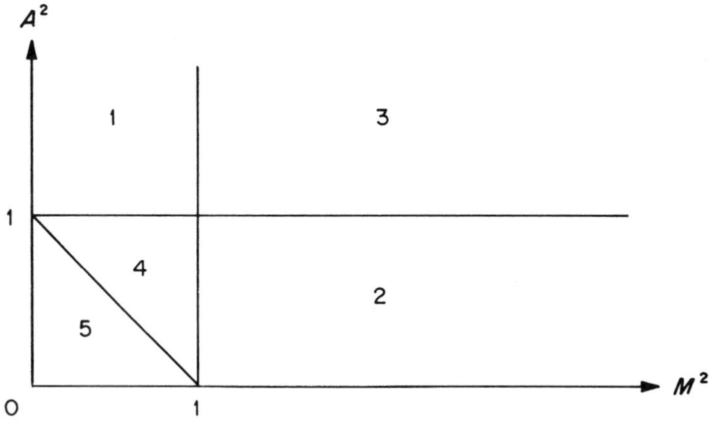

FIG. 5.1.1.a. Regions 1–5 for A^2, M^2 coordinates.

We can then classify the flows according to different criteria; for example,

Elliptic flows: regions 1, 2, 5
Hyperbolic flows: regions 3, 4
Sub-Alfvénic flows: regions 2, 4, 5
Super-Alfvénic flows: regions 1, 3

In a given flow, the specific entropy being constant, the Alfvén number and Mach number are not independent. For a polytropic gas, we have, for example,

$$A^2 = A_0^2 \frac{\rho_0}{\rho} = A_0^2 \left\{ 1 - \frac{\gamma - 1}{2} \frac{V^2}{c_0^2} \right\}^{-1(\gamma-1)}$$

$$M^2 = \frac{V^2}{c^2} = \frac{V^2}{c_0^2} \left\{ 1 - \frac{\gamma - 1}{2} \frac{V^2}{c_0^2} \right\}^{-1}$$

Since Bernoulli's equation still holds in such a flow, we deduce that

$$\left(\frac{A^2}{A_0^{\,2}}\right)^{\gamma-1} = 1 + \frac{\gamma-1}{2} M^2 \qquad (5.1.1\text{-}7)$$

Note that, in a flow where the magnetic field is aligned with the velocity, this quantity does not change across a stationary shock (Property 9, Section 3.1.1).

5.1.2. Flows with a Transverse Magnetic Field

As in the previous sections, we are still assuming the fluid to be inviscid and a perfect conductor of electricity, but we neglect thermal conduction. The equations of magnetofluiddynamics admit solutions for which **H** and **V** have the simple forms

$$\mathbf{H} = \mathbf{z}\, H(t, x, y)$$
$$\mathbf{V} = \mathbf{x}\, u(t, x, y) + \mathbf{y}\, v(t, x, y)$$

For this motion, the general equations are written

$$\begin{aligned} \rho\left(\frac{\partial \mathbf{V}}{\partial t} + \mathbf{V}\cdot\nabla\mathbf{V}\right) + \nabla\left(p + \frac{\mu H^2}{2}\right) &= 0 \\ \frac{\partial \rho}{\partial t} + \nabla\cdot(\rho\mathbf{V}) &= 0 \\ \frac{\partial H}{\partial t} + \nabla\cdot(H\mathbf{V}) &= 0 \\ \frac{\partial S}{\partial t} + \mathbf{V}\cdot\nabla\mathbf{S} &= 0 \end{aligned} \qquad (5.1.2\text{-}1)$$

The third equation shows that $(\partial\mathbf{H}/\partial t) + \nabla \times \mathbf{E} = 0$, where $\mathbf{E} = -\mathbf{V}\times\mathbf{H}$ can be replaced by the relation

$$\frac{\partial}{\partial t}\left(\frac{H}{\rho}\right) + \mathbf{V}\cdot\nabla\left(\frac{H}{\rho}\right) = 0 \qquad (5.1.2\text{-}2)$$

Then we have

$$\begin{aligned} &\frac{\partial}{\partial t}\left(\frac{H}{\rho}\right) + \mathbf{V}\cdot\nabla\left(\frac{H}{\rho}\right) \\ &\quad = \frac{1}{\rho}\left\{\frac{\partial H}{\partial t} + \mathbf{V}\cdot\nabla H\right\} - \frac{H}{\rho^2}\left\{\frac{\partial \rho}{\partial t} + \mathbf{V}\cdot\nabla\rho\right\} \end{aligned}$$

$$= \frac{1}{\rho}\{-\nabla \cdot (H\mathbf{V}) + \mathbf{V} \cdot \nabla H\} - \frac{H}{\rho^2}\{-\nabla \cdot (\rho\mathbf{V}) + \mathbf{V} \cdot \nabla\rho\}$$

$$\equiv \frac{1}{\rho}\{-H\nabla \cdot \mathbf{V}\} + \frac{H}{\rho^2}\{\rho\nabla \cdot \mathbf{V}\} = 0$$

We introduce the new quantities

$$p^* = p + \tfrac{1}{2}\mu H^2, \qquad s^* = H/\rho, \qquad \epsilon_i^* = \epsilon_i + (\mu H^2/\rho)$$

which enable us to write (5.1.2-1) and (5.1.2-2) in the equivalent form

$$\begin{aligned} \rho\left(\frac{\partial \mathbf{V}}{\partial t} + \mathbf{V} \cdot \nabla\mathbf{V}\right) + \nabla p^* &= 0 \\ \frac{\partial \rho}{\partial t} + \nabla \cdot (\rho\mathbf{V}) &= 0 \\ \frac{\partial s^*}{\partial t} + \mathbf{V} \cdot \nabla s^* &= 0 \\ \rho\left(\frac{\partial \epsilon_i^*}{\partial t} + \mathbf{V} \cdot \nabla\epsilon_i\right) + p^*\nabla \cdot \mathbf{V} &= 0 \end{aligned} \tag{5.1.2-3}$$

The first three equations replace the first two equations of (5.1.2-1) and Eq. (5.1.2-2), respectively. The fourth one replaces the fourth of Eq. (5.1.2-1), which can be written as

$$\rho\left(\frac{\partial \epsilon_i}{\partial t} + \mathbf{V} \cdot \nabla\epsilon_i\right) + p\nabla \cdot \mathbf{V} = 0$$

Then we also have

$$\rho\left\{\frac{\partial}{\partial t}\left(\frac{\mu H^2}{2\rho}\right) + \mathbf{V} \cdot \nabla\left(\frac{\mu H^2}{2\rho}\right)\right\} + \frac{\mu H^2}{2}\nabla \cdot \mathbf{V} = 0$$

Using the second and third of Eqs. (5.1.2-1), we can deduce the fourth of Eqs. (5.1.2-3).

The first two and the fourth of Eqs. (5.1.2-3) represent the aerodynamic flow of a fictitious fluid defined by $\mathbf{V}$, p^*, and ρ, provided that ϵ_i^* is a function of p^* and ρ.

We have

$$\epsilon_i^* = \epsilon_i(p, \rho) + (\mu H^2/2\rho)$$

$$\epsilon_i^* = \epsilon_i\{p^* - \tfrac{1}{2}\mu\rho^2 s^{*2}, \rho\} + \tfrac{1}{2}\mu\rho s^{*2}$$

If we are to express ϵ_i^* as a function of p^* and ρ alone, we must have p^*, ρ, and s^* related in some way; this would be the case if, for example, $s^* = f(S)$. The third of Eqs. (5.1.2-3) is then automatically satisfied because it reduces to $\partial S/\partial t + \mathbf{V} \cdot \nabla \mathbf{S} = 0$.

Let us assume that a relation $s^* = f(S)$ does exist; the equation of state of the fictitious fluid is then written

$$p^*(\rho, s^*) = p(\rho, S) + \tfrac{1}{2}\mu\rho^2 s^{*2}$$

where S has to be expressed in terms of s^*. The speed of propagation of magnetoacoustic waves will be given by

$$a^2 = \left(\frac{\partial p^*}{\partial \rho}\right)_{s^*} = \left(\frac{\partial p}{\partial \rho}\right)_S + \frac{\mu H^2}{2}$$

There are several cases for which some relationship of the form $s^* = f(S)$ exists.

(i) *One-Dimensional Flows.* We have

$$\frac{\partial S}{\partial t} + u\frac{\partial S}{\partial x} = 0, \qquad \frac{\partial s^*}{\partial t} + u\frac{\partial s^*}{\partial x} = 0$$

Then

$$\frac{D(S, s^*)}{D(t, x)} = 0, \qquad \text{or} \qquad s^* = f(S)$$

(ii) *Plane Steady Flows.* We have

$$u\frac{\partial S}{\partial x} + v\frac{\partial S}{\partial y} = 0, \qquad u\frac{\partial s^*}{\partial x} + v\frac{\partial s^*}{\partial y} = 0$$

Then

$$\frac{D(S, s^*)}{D(x, y)} = 0, \qquad \text{or} \qquad s^* = f(S)$$

(iii) The initial value $s^*(0, x, y)$ is constant; therefore, the function $s^*(t, x, y)$ is constant for all time.

5.1.3. Viscous Flows

We now consider the fluid to be viscous and a perfect electrical conductor, but thermal conduction is neglected. The equations of motion can be written

$$\begin{gathered} \rho_0 \mathbf{V} \cdot \nabla \mathbf{V} + \nabla p = (\nabla \times \mathbf{H}) \times \mu \mathbf{H} + 2\mu_1 \nabla \cdot \mathbf{D}_1 \\ \nabla \cdot \mathbf{V} = 0, \qquad \nabla \times (\mathbf{V} \times \mathbf{H}) = 0 \end{gathered} \tag{5.1.3-1}$$

The tensor $\mathbf{D}_1$ is the deviator of the symmetric tensor associated with the gradient of the velocity; and, since the divergence of this velocity is zero, we find that $2\nabla \cdot \mathbf{D}_1 = \nabla^2 \mathbf{V}$.

Equations (5.1.3-1) admit solutions for which the magnetic field and velocity are aligned; but the difference from the motions studied in Section 5.1.1 is due to the presence of viscosity and the absence of compressibility.

We write

$$\mathbf{H} = \beta(\rho_0/\mu)^{1/2}\,\mathbf{V},$$

so that β^{-1} represents the Alfvén number. The density being constant, the equation $\nabla \cdot \mathbf{H} = 0$, which corresponds to certain initial conditions, gives

$$\beta\nabla \cdot \mathbf{V} + \mathbf{V} \cdot \nabla\beta = \mathbf{V} \cdot \nabla\beta = 0$$

The Alfvén number is then constant on each streamline. If the velocity and the magnetic field are uniform at infinity and if all streamlines go through infinity, which we assume to be the case, the Alfvén number is then constant throughout.

Using the identity

$$(\nabla \times \mathbf{H}) \times \mathbf{H} \equiv \mathbf{H} \cdot \nabla\mathbf{H} - \nabla(H^2/2)$$

we can write Eqs. (5.1.3-1) in the form

$$\rho_0(1 - \beta^2)\,\mathbf{V} \cdot \nabla\mathbf{V} + \nabla(p + \tfrac{1}{2}\mu H^2) = \mu_1\nabla^2\mathbf{V}, \qquad \nabla \cdot \mathbf{V} = 0 \quad (5.1.3\text{-}2)$$

When $\beta < 1$ (super-Alfvénic flow), Eqs. (5.1.3-2) represent the classical Navier–Stokes equation for a fluid having a density equal to $\rho_0(1 - \beta^2)$.

When $\beta > 1$ (sub-Alfvénic flows), we can write

$$\mathbf{V}^* = -\mathbf{V}, \qquad \rho_0{}^* = (\beta^2 - 1)\,\rho_0\,, \qquad P^* = -(p + \tfrac{1}{2}\mu H^2)$$

and again recover the classical Navier–Stokes equations for a fictitious fluid deduced from the physical flow by multiplying the velocity by -1. This negative factor explains the fact that in sub-Alfvénic flows we find wakes oriented upstream rather than downstream as in classical fluid mechanics.

5.2. Flow Past a Semi-infinite Flat Plate

5.2.1. Boundary Layer Equations

The flow over a semi-infinite or finite flat plate is a fundamental problem of fluid mechanics since it provides a simple model for the study of the influence of viscosity on flow near a solid boundary; this is the boundary layer problem.

The equations of motion of an electrically conducting incompressible viscous flow are written in the form

$$\frac{\partial \mathbf{V}}{\partial t} + \mathbf{V} \cdot (\nabla \cdot \mathbf{V}) + \frac{\nabla p}{\rho_0} = \nu \nabla^2 \mathbf{V} + \frac{(\nabla \times \mathbf{H}) \times \mu \mathbf{H}}{\rho_0}$$

$$\nabla \cdot \mathbf{V} = 0, \qquad \nabla \cdot \mathbf{H} = 0$$

$$\nabla \times \mathbf{H} = \sigma (\mathbf{E} + \mathbf{V} \times \mu \mathbf{H}) \tag{5.2.1-1}$$

$$\nabla \times \mathbf{E} = -\frac{\partial \mu \mathbf{H}}{\partial t}$$

It is convenient to introduce some reference magnetic field $\mathbf{H}_\infty$ and velocity $\mathbf{V}_\infty$ and to write the above equations in dimensionless form. For this purpose, we multiply the time by V_∞^2/ν, geometrical lengths by V_∞/ν, and divide $\mathbf{H}$, $\mathbf{E}$, $\mathbf{V}$, and p by H_∞, $\mu H_\infty V_\infty$, V_∞, and $\rho_0 V_\infty^2$, respectively. Keeping the same notation for simplicity, we obtain the dimensionless form of Eqs. (5.2.1-1)

$$\frac{\partial \mathbf{V}}{\partial t} + \mathbf{V} \cdot (\nabla \cdot \mathbf{V}) + \nabla p = \nabla^2 \mathbf{V} + \beta (\nabla \times \mathbf{H}) \times \mathbf{H}$$

$$\nabla \cdot \mathbf{V} = 0, \qquad \nabla \cdot \mathbf{H} = 0$$

$$\nabla \times \mathbf{H} = \epsilon \{\mathbf{E} + \mathbf{V} \times \mathbf{H}\} \tag{5.2.1-2}$$

$$\nabla \times \mathbf{E} = -\frac{\partial \mathbf{H}}{\partial t}$$

where

$$\beta = \mu H_0^2 / \rho_0 V_0^2, \qquad \epsilon = \sigma \mu \nu$$

The problem involves only two dimensionless parameters, β (the inverse square of the Alfvén number) and ϵ. In the case of plane flows and flows with axial symmetry, the vector fields $\mathbf{H}$ and $\mathbf{V}$ are derivatives of a vector potential. We will consider plane flows: $(\partial/\partial z) = 0$; also, $H_z = 0$, $V_z = 0$. We then have

$$\mathbf{H} = \nabla \times (\mathbf{z}\, A(t, x, y)), \qquad \mathbf{V} = \nabla \times (\mathbf{z}\, \Psi(t, x, y))$$

The electric field is parallel to z and the second of Eqs. (5.2.1-1) becomes $E + \partial A/\partial t = \text{const}$. This constant can be taken as zero because we can always add to A some linear function of t. Equating the curls of the right- and left-hand sides of Eq. (5.2.1-2), we can eliminate the pressure and obtain the first of the following equations, the third of Eqs. (5.2.1-2) yielding the second:

$$\Delta\Delta\Psi - \frac{\partial\Psi}{\partial y}\Delta\left(\frac{\partial\Psi}{\partial x}\right) + \frac{\partial\Psi}{\partial x}\Delta\left(\frac{\partial\Psi}{\partial y}\right) - \Delta\left(\frac{\partial\Psi}{\partial t}\right)$$

$$+ \beta\left\{\frac{\partial A}{\partial y}\Delta\left(\frac{A}{\partial x}\right) - \frac{\partial A}{\partial x}\Delta\left(\frac{\partial A}{\partial y}\right)\right\} = 0 \qquad (5.2.1\text{-}3)$$

$$\Delta A + \epsilon\left\{\frac{\partial\Psi}{\partial x}\frac{\partial A}{\partial y} - \frac{\partial\Psi}{\partial y}\frac{\partial A}{\partial x} - \frac{\partial A}{\partial t}\right\} = 0$$

where $\Delta = \partial^2/\partial x^2 + \partial^2/\partial y^2$.

We will henceforth limit attention to steady flows, i.e., $\partial/\partial t = 0$, and make the change of variables defined by the complex relation

$$\xi + i\eta = (x + iy)^{1/2}$$

The flat plate forms the semiaxis $y = 0$, $x \geqslant 0$ and has as its equation $\eta = 0$. The problem consists in finding a solution of Eqs. (5.2.1-3) satisfying the boundary conditions:

On the flat plate, $\quad u = \partial\Psi/\partial y = 0, \quad v = -\partial\Psi/\partial x = 0,$

$\quad H_y = -\partial A/\partial x = 0.$

At infinity, $\quad u = \partial\Psi/\partial y = 1, \quad v = -\partial\Psi/\partial x = 0,$

$\quad H_x = \partial A/\partial y = 1, \quad H_y = -\partial A/\partial x = 0.$

Since we cannot find an exact solution, we will look for an approximate solution of this problem using the functions $\Psi(x, y)$ and $A(x, y)$ in the form

$$\Psi(x, y) = \xi f(\eta), \qquad A(x, y) = \xi g(\eta) \qquad (5.2.1\text{-}4)$$

Substituting these values in Eqs. (5.2.1-3), we find the two relations

$$\frac{d}{d\eta}\{f''' + ff'' - \beta g g''\} = \frac{2\eta F(\eta)}{\xi^2 + \eta^2}$$

$$g'' + \epsilon(fg' - f'g) = 0$$

with

$$F(\eta) \equiv 2f'''' + f''(\eta f' + f) - \beta\{2g'''' + g''(\eta' g + g)\}$$

We have

$$u = \frac{\eta f + \xi^2 f'}{2(\xi^2 + \eta^2)}, \qquad v = -\frac{\xi(f - \eta f')}{2(\xi^2 + \eta^2)}$$

$$H_x = \frac{\eta g + \xi^2 g'}{2(\xi^2 + \eta^2)}, \qquad H_y = -\frac{\xi(g - \eta g')}{2(\xi^2 + \eta^2)}$$

The first of the relations between f and g can only be satisfied if both sides of the equation vanish, but in this case the second relation is not satisfied. However, when moving on a line $\eta =$ const, we let ξ increase indefinitely (Fig. 5.2.1.a); the quotient $2\eta F(\eta)/(\xi^2 + \eta^2)$ then tends to zero; this means that, for from the leading edge, we can write,

$$f''' + ff'' - \beta g g'' = k = \text{const}$$

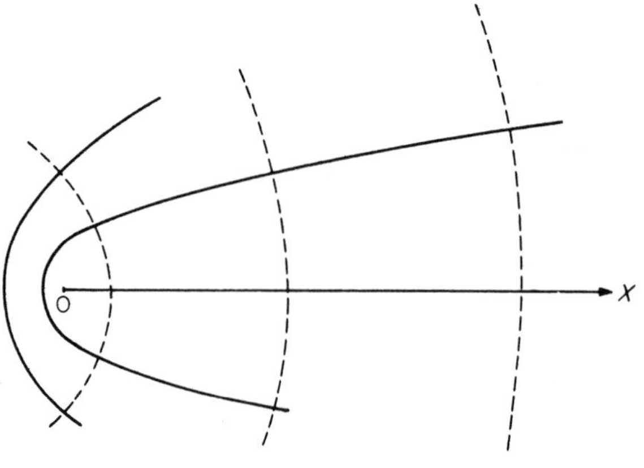

FIG. 5.2.1.a. Curves: dashed lines, $\xi =$ const; solid lines, $\eta =$ const.

Under the same conditions, we have

$$u \sim \tfrac{1}{2}f'(\eta), \qquad v \sim -[f(\eta) + \eta f'(\eta)]/2\xi$$

$$H_x \sim \tfrac{1}{2}g'(\eta), \qquad H_y \sim -[g(\eta) + \eta g'(\eta)]/2\xi$$

The boundary conditions then become:

On the plate, $f(0) = f'(0) = 0, \quad g(0) = 0.$

At infinity, $f'(\infty) = 2, \quad g'(\infty) = 2.$

Assuming that the derivatives $f'(\eta)$ and $g'(\eta)$ tend to infinity in a regular fashion, we have $f''(\infty) = f'''(\infty) = g''(\infty) = 0$; hence, $k = 0$ and the functions $f(\eta)$ and $g(\eta)$ are solutions of the equations

$$\begin{aligned} f''' + ff'' - \beta g g'' &= 0 \\ g'' + \epsilon(fg' - f'g) &= 0 \end{aligned} \tag{5.2.1-5}$$

Equations (5.2.1-5) are the boundary layer equations in magneto-fluiddynamics. It is possible to express these equations in a more symmetrical form; for this purpose, take the derivative of the first equation and add the boundary condition $g''(0) = 0$. We then write

$$\Phi(\eta) = f(\eta) - g(\eta)\sqrt{\beta}, \qquad \Psi_0(\eta) = f(\eta) + g(\eta)\sqrt{\beta}$$

The functions $\Phi(\eta)$ and $\Psi_0(\eta)$ are then determined by the following equations and boundary conditions:

$$\begin{aligned} \Phi''' + \tfrac{1}{2}(1+\epsilon)\,\Psi_0\Phi'' + \tfrac{1}{2}(1-\epsilon)\,\Phi\Psi_0'' &= 0 \\ \Psi_0''' + \tfrac{1}{2}(1+\epsilon)\,\Phi\Psi_0'' + \tfrac{1}{2}(1-\epsilon)\,\Psi_0\Phi'' &= 0 \end{aligned} \tag{5.2.1-6}$$

$$\Phi(0) = \Psi_0(0) = 0 \tag{5.2.1-7a}$$

$$\Phi'(0) + \Psi_0'(0) = 0 \tag{5.2.1-7b}$$

$$\Phi''(0) - \Psi_0''(0) = 0 \tag{5.2.1-7c}$$

$$\Phi'(\infty) = 2(1 - \sqrt{\beta}) \tag{5.2.1-7d}$$

$$\Psi_0'(\infty) = 2(1 + \sqrt{\beta}) \tag{5.2.1-7e}$$

Since β is strictly positive, each solution of this problem corresponds to a solution of the previous one, and conversely. But the new conditions

$$\begin{aligned}\Phi(0) &= \Psi_0(0) = 0\\ \Phi'(0) &= \Psi_0'(0) = A\\ \Phi''(0) &= \Psi_0''(0) = B\end{aligned} \tag{5.2.1-7$'$}$$

determine a unique solution of Eq. (5.2.1-6), and this solution exists in the neighborhood of the origin. Hence, if the conditions (5.2.1-7a–e) determine a solution for some values of $\beta \geqslant 0$, this solution is given by the condition (5.2.1-7′) with an appropriate choice of A and $B \neq 0$, for, if $B = 0$, the solution is $\Phi = -A\eta$, $\Psi_0 = A\eta$, which does not satisfy the conditions (5.2.1-7d,e).

We then call $\varphi(\eta, \alpha)$, $\Psi_1(\eta, \alpha)$ the solution of Eqs. (5.2.1-6) which satisfies the conditions

$$\begin{aligned}\varphi(0, \alpha) &= \Psi_1(0, \alpha) = 0\\ -\varphi'(0, \alpha) &= \Psi_1'(0, \alpha) = \alpha\\ \varphi''(0, \alpha) &= \Psi_1''(0, \alpha) = 1\end{aligned} \tag{5.2.1-7$''$}$$

where α is a real number and where the primes mean derivatives with respect to η.

For real λ, the functions

$$\Phi(\eta) = \lambda\varphi(\lambda\eta, \alpha), \qquad \Psi_0(\eta) = \lambda\Psi_1(\lambda\eta, \alpha)$$

satisfy the system (5.2.1-6). If $B \neq 0$, there exists a unique pair of values λ, α such that conditions (5.2.1-7a–e) are satisfied:

$$\lambda = B^{1/3}, \qquad \alpha = AB^{-2/3}$$

Thus, each solution of Eqs. (5.2.1-7′) with $B \neq 0$ can be expressed in this form. We then have

$$\begin{aligned}\lambda^2\varphi'(\infty, \alpha) &= 2(1 - \sqrt{\beta})\\ \lambda^2\Psi_1'(\infty, \alpha) &= 2(1 + \sqrt{\beta})\end{aligned}$$

Conversely, if there exist numbers λ and α satisfying these relations, then $\lambda\varphi(\lambda\eta, \alpha)$, $\lambda\Psi_1(\lambda\eta, \alpha)$ is a solution of the system (5.2.1-6) satisfying Eqs. (5.2.1-7a–e).

Theorem. For $\beta \geqslant 0$, the boundary value problem (5.2.1-6)–(5.2.1-7) has a solution if, and only if, there exists a number α such that $\varphi(\eta, \alpha)$ and $\Psi_1(\eta, \alpha)$ are defined for all $\eta \geqslant 0$, and $\varphi'(\infty, \alpha)$ and $\Psi_1'(\infty, \alpha)$ are finite and $\varphi'(\infty, \alpha)$ is positive,

$$\frac{\varphi'(\infty, \alpha)}{\Psi_1'(\infty, \alpha)} = \frac{1 - \sqrt{\beta}}{1 + \sqrt{\beta}} \tag{5.2.1-8}$$

For each α satisfying the above conditions, the solution can be written

$$\Phi(\eta) = \lambda\varphi(\lambda\eta, \alpha) \qquad \Psi_0(\eta) = \lambda\Psi_1(\lambda\eta, \alpha)$$

with

$$\lambda = \{2(1 + \sqrt{\beta})/\Psi_1'(\infty, \alpha)\}^{1/2}$$

If there exists a single value of α satisfying the above conditions, the boundary value problem (5.2.1-6)–(5.2.1-7) has a unique solution. The functions $\Phi = \varphi$ and $\Psi_0 = \Psi_1$ by definition make up one solution of system (5.2.1-6), but so do the functions $\Phi = \Psi_1$ and $\Psi_0 = \varphi$. We can then deduce from conditions (5.2.1-7″)

$$\varphi(\eta, -\alpha) \equiv \Psi_1(\eta, \alpha)$$

for each real α and for each η for which these functions exist.

5.2.2. Uniqueness of the Solution for $\epsilon = 1$

When $\epsilon = 1$, Eqs. (5.2.1-6) are (writing Ψ in place of Ψ_0)

$$\begin{aligned} \varphi''''(\eta) + \Psi(\eta)\,\varphi''(\eta) &= 0 \\ \Psi''''(\eta) + \varphi(\eta)\,\Psi''(\eta) &= 0 \end{aligned} \tag{5.2.2-1}$$

Taking account of the boundary conditions (5.2.1-7″), we deduce

$$\begin{aligned} \varphi''(\eta) &= \exp\left\{-\int_0^\eta \Psi(\tau)\,d\tau\right\} \\ \Psi''(\eta) &= \exp\left\{-\int_0^\eta \varphi(\tau)\,d\tau\right\} \end{aligned} \tag{5.2.2-2}$$

$$\begin{aligned} \varphi'(\eta) &= -\alpha + \int_0^\eta \exp\left\{-\int_0^s \Psi(\tau)\,d\tau\right\} ds \\ \Psi'(\eta) &= \alpha + \int_0^\eta \exp\left\{-\int_0^s \varphi(\tau)\,d\tau\right\} ds \end{aligned} \tag{5.2.2-3}$$

$$\varphi(\eta) = -\alpha\,\eta + \int_0^\eta (\eta - s) \exp\left\{-\int_0^s \Psi(\tau)\,d\tau\right\} ds$$
$$\Psi(\eta) = \alpha\,\eta + \int_0^\eta (\eta - s) \exp\left\{-\int_0^s \varphi(\tau)\,d\tau\right\} ds \qquad (5.2.2\text{-}4)$$

We will first prove that the solution of Eq. (5.2.1-6) with conditions (5.2.1-7″) can be continued along the whole semiaxis $\eta \geqslant 0$. The functions $\varphi(\eta)$ and $\Psi(\eta)$ are convex with positive second derivatives. If they are defined for $0 \leqslant \eta < \eta_0$, their limits as $\eta \to \eta_0$ are either finite or positive infinite. If $\varphi(\eta_0 - 0) = +\infty$, Eqs. (5.2.2-4) are incompatible; hence, the limits $\varphi(\eta_0 - 0)$ and $\Psi(\eta_0 - 0)$ are finite. The functions φ', Ψ', φ'', and Ψ'' determined from Eqs. (5.2.2-3) and (5.2.2-2) also have finite limits for $\eta \to \eta_0$ and the solution can be continued beyond η_0 up to η_1, etc. The values η_0, η_1, η_2,... do not have a limit $\bar{\eta}$, since, for at $\bar{\eta}$, the values of $\varphi, \varphi', \varphi'', \Psi, \Psi'$, and Ψ'' would still be finite and could not be extended beyond $\bar{\eta}$, which contradicts Cauchy's theorem.

Hence, the functions $\varphi(\eta, \alpha)$, $\Psi(\eta, \alpha)$, $\varphi'(\eta, \alpha)$, $\Psi'(\eta, \alpha)$ exist and are continuous for all $\eta \geqslant 0$ and all real α; and, by a classical theorem of analysis, these functions are continuous functions of α because they depend continuously on the initial data.

The function $\varphi'(\eta, \alpha)$ and $\Psi'(\eta, \alpha)$ are then monatonically increasing functions of η, and consequently the limits $\varphi'(\infty, \alpha)$ and $\Psi'(\infty, \alpha)$ exist. To be able to discuss Eqs. (5.2.1-8), we will study the behavior of those limits as a function of α. The remark made at the end of Section 5.2.1 justifies considering only the case $\alpha \geqslant 0$.

Theorem 1. For each $\alpha \geqslant 0$, $\varphi'(\infty, \alpha)$ is finite; $\varphi'(\infty, \alpha) > 0 \to \Psi'(\infty, \alpha)$ finite >0, $\varphi'(\infty, \alpha) \leqslant 0 \to \Psi'(\infty, \alpha)$ infinite.

By Eqs. (5.2.2-3), if $\alpha \geqslant 0$, we have $\Psi'(1) > 0$, and, since $\Psi'(\eta)$ is monotonic, $\Psi'(\eta) > \Psi'(1)$ for $\eta > 1$. Hence,

$$\Psi(\eta) > \Psi(1) + (\eta - 1)\,\Psi'(1)$$

for $\eta > 1$, implies the convergence of the integral in the first of Eqs. (5.2.2-3). Consequently, $\varphi'(\infty, \alpha)$ is finite.

If we have $\varphi'(\infty, \alpha) > 0$, then, for some $\eta_1 > 0$, $\varphi'(\eta) > 0$, and $\varphi(\eta) > \varphi(\eta_1) + (\eta - \eta_1)\,\varphi'(\eta_1)$ for $\eta > \eta_1$. From the second of Eqs. (5.2.2-3), $\varphi'(\infty, \alpha)$ is finite; it is also positive.

If we have $\varphi'(\infty, \alpha) \leqslant 0$, then, by the monotonic property of

$\varphi'(\eta)$, $\varphi'(\eta) < 0$ for all $\eta \geqslant 0$. Since $\varphi(0) = 0$, $\varphi(\eta)$ is negative for all $\eta > 0$; it follows from Eq. (5.2.2-3) that $\Psi'(\infty, \alpha)$ is infinite.

Theorem 2. There exist two positive numbers α_0 and α_1 such that $0 \leqslant \alpha < \alpha_0 \rightarrow \varphi'(\infty, \alpha) > 0$ and $\alpha_1 < \alpha \rightarrow \varphi'(\infty, \alpha) < 0$.

We have, as a matter of fact, $\varphi'(1, 0) > 0$, so that, by the continuity of $\varphi'(\eta, \alpha)$, there exists a number $\alpha_0 > 0$ such that $\varphi'(1, \alpha)$ for $0 \leqslant \alpha < \alpha_0$. Knowing that $\varphi'(\eta, \alpha)$ is monotonic, we can deduce that $\varphi'(\infty, \alpha)$ is positive for $0 \leqslant \alpha < \alpha_0$.

We also have $\Psi(\eta, \alpha) > \alpha\eta$ for $\eta > 0$; hence, for $\alpha > 0$,

$$\varphi'(\infty, \alpha) < -\alpha + \int_0^\infty \exp(-\alpha s^2/2)\, ds = -\alpha + (\pi/2\alpha)^{1/2}$$

It follows that $\alpha^3 > \pi/2 \rightarrow \varphi'(\infty, \alpha) < 0$.

Theorem 3. If $0 \leqslant \alpha_1 < \alpha_2$, then, for each $\eta > 0$,

$$\varphi(\eta, \alpha_1) > \varphi(\eta, \alpha_2), \qquad \Psi(\eta, \alpha_1) < \Psi(\eta, \alpha_2)$$

$$\varphi'(\eta, \alpha_1) > \varphi'(\eta, \alpha_2) + (\alpha_2 - \alpha_1) > \varphi'(\eta, \alpha_2)$$

$$\Psi'(\eta, \alpha_1) < \Psi'(\eta, \alpha_2) - (\alpha_2 - \alpha_1) < \Psi'(\eta, \alpha_2)$$

From the initial conditions, we have $\varphi(\eta, \alpha_1) > \varphi(\eta, \alpha_2)$ and $\Psi(\eta, \alpha_1) < \Psi(\eta, \alpha_2)$ in some interval $0 < \eta < \eta_0$. It follows from Eqs. (5.2.2-4) that, if these inequalities are valid in an open interval $0 < \eta < \eta_1$, they are also valid for all $\eta > 0$. The inequalities for φ' and Ψ' then follow from Eqs. (5.2.2-3).

Theorem 4. There exists a number $\alpha^* > 0$ such that

$$0 \leqslant \alpha < \alpha^* \rightarrow \varphi'(\infty, \alpha) > 0 \qquad \text{and} \qquad \alpha^* < \alpha \rightarrow \varphi'(\infty, \alpha) < 0.$$

We denote by α^* the upper bound of the numbers $\alpha_0 > 0$, such that $0 \leqslant \alpha < \alpha_0 \rightarrow \varphi'(\infty, \alpha) > 0$. By Theorem 2, α^* is bounded. By the definition of α^*, we know that $0 \leqslant \alpha < \alpha^* \rightarrow \varphi'(\infty, \alpha) > 0$.

On the other hand, if $\alpha > \alpha^*$, there exists a number α_1 in the interval $\alpha^* < \alpha_1 < \alpha$ such that $\varphi'(\infty, \alpha_1) < 0$. From Theorem 3, we have $\varphi'(\infty, \alpha) < \varphi'(\infty, \alpha_1) < 0$.

It follows from Theorems 2 and 4 that

$$0 \leqslant \alpha < \alpha^* \rightarrow \Psi'(\infty, \alpha) \qquad \text{finite and positive}$$

$$\alpha^* < \alpha \rightarrow \Psi'(\infty, \alpha) \qquad \text{infinite}$$

Theorem 5. For all $\alpha \geqslant 0$, $\varphi'(\infty, \alpha)$ is a continuous function of α. For $0 \leqslant \alpha < \alpha^*$, $\Psi'(\infty, \alpha)$ is also a continuous function of α; and, for $\alpha \to \alpha^*$, $\Psi'(\infty, \alpha) \to \infty$.

We consider a sequence $\alpha_n \geqslant 0$, $n = 1, 2, \ldots$, which tends to α_0 ; we will prove that $\varphi'(\infty, \alpha_n) \to \varphi'(\infty, \alpha_0)$.

The continuity of $\Psi(\eta, \alpha)$ as a function of α (for η finite and any α) implies

$$\lim_{n\to\infty} \exp\left\{-\int_0^s \Psi(\tau, \alpha_n)\, d\tau\right\} = \exp\left\{-\int_0^s \Psi(\tau, \alpha_0)\, d\tau\right\}$$

We can also write

$$\begin{aligned}\Bigg| \int_0^\infty & \exp\left\{-\int_0^s \Psi(\tau, \alpha_n)\, d\tau\right\} ds - \int_0^\infty \exp\left\{-\int_0^s \Psi(\tau, \alpha_0)\, d\tau\right\} ds \Bigg| \\ &\leqslant \left| \int_0^S \exp\left\{-\int_0^s \Psi(\tau, \alpha_n)\, d\tau\right\} ds - \int_0^S \exp\left\{-\int_0^s \Psi(\tau, \alpha_0)\, d\tau\right\} ds \right| \\ &\quad + \left| \int_S^\infty \exp\left\{-\int_0^s \Psi(\tau, \alpha_n)\, d\tau\right\} ds \right| + \left| \int_S^\infty \exp\left\{-\int_0^s \Psi(\tau, \alpha_0)\, d\tau\right\} ds \right|\end{aligned}$$

The first term on the right-hand side of this inequality can be as small as we please because S is bounded. This is also true for each of the other terms of the right-hand side; in fact, by Theorem 3, we have

$$0 < \exp\left\{-\int_0^s \Psi(\tau, \alpha)\, d\tau\right\} < \exp\left\{-\int_0^s \Psi(\tau, 0)\right\} d\tau$$

and, from Theorem 1,

$$\varphi'(\infty, 0) = \int_0^\infty \exp\left\{-\int_0^s \Psi(\tau, 0)\, d\tau\right\} ds < \infty$$

It follows that $\varphi'(\infty, \alpha)$ is a continuous function of α.

We now assume that we have $0 \leqslant \alpha_n < \alpha^*$, and that, for $n \to \infty$, $\alpha_n \to \alpha_0$. There then exists a number b such that $\alpha_n \leqslant b < \alpha^*$, $n = 0, 1, \ldots$, and an analogous proof starting from the inequality $\varphi(\eta, \alpha_n) \geqslant \varphi(\eta, b)$ shows that $\Psi'(\infty, \alpha)$ is a continuous function of α.

To prove the last part of the theorem, we assume that we have $0 < \alpha_n < \alpha^*$, $n = 1, 2, \ldots$, and $\alpha_n \to \alpha^*$. It follows from Theorem 4 and the continuity of $\varphi'(\infty, \alpha)$ that $\varphi'(\infty, \alpha^*) = 0$. Hence, by Theorem 1, $\Psi'(\infty, \alpha^*) = \infty$; but, knowing that $\Psi'(\eta)$ is monotonic,

$$\Psi'(\infty, \alpha_n) > \Psi'(\eta, \alpha_n)$$

If the limit of $\Psi'(\infty, \alpha_n)$ as $n \to \infty$ were not infinite, we would have $\Psi'(\infty, \alpha) < B$ for $n > N$; hence $\Psi'(\infty, \alpha^*) < 2B$. Since the limit B is independent of η, this contradicts the fact that $\Psi'(\infty, \alpha^*) = \infty$. This proves the last part of the theorem.

Theorem 6. For $0 \leqslant \alpha < \alpha^*$, the function

$$\rho(\alpha) = \varphi'(\infty, \alpha)/\Psi'(\infty, \alpha)$$

is continuous, monotonic, and strictly decreasing (Fig. 5.2.2a). On the other hand, $\rho(0) = 1$ and $\rho(\alpha^* - 0) = 0$.

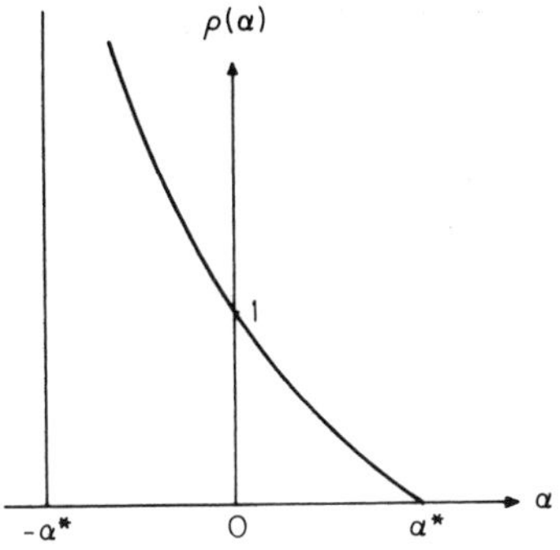

FIG. 5.2.2.a. The function $\rho(\alpha)$.

This theorem follows from the five preceding ones. The value $\rho(0) = 1$ follows from the fact that $\varphi(\eta, -\alpha) \equiv \Psi(\eta, \alpha)$ leads to $\varphi(\eta, 0) \equiv \Psi(\eta, 0)$. Since the ratio $(1 - \sqrt{\beta})/(1 + \sqrt{\beta})$ is continuous, monotonic, and strictly decreasing in the interval $0 \leqslant \beta < 1$, and has the value 1 for $\beta = 0$ and 0 for $\beta = 1$ (Fig. 5.2.2.b), Eq. (5.2.1-8) determines α uniquely as a function of β; this function is continuous, monotonic, and strictly decreasing in the interval $0 \leqslant \beta < 1$, and $\alpha = 0$ for $\beta = 0$, while $\alpha \to \alpha^*$ when $\beta \to 1$.

Theorem 7. The boundary value problem (5.2.1-6)–(5.2.1-7) admits, for $\epsilon = 1$, one and only one solution if $0 \leqslant \beta < 1$, and no solutions if $\beta > 1$.

The first case (uniqueness of the solution) corresponds to super-Alfvénic flows, the second (nonexistence of solution) corresponds to sub-Alfvénic flows.

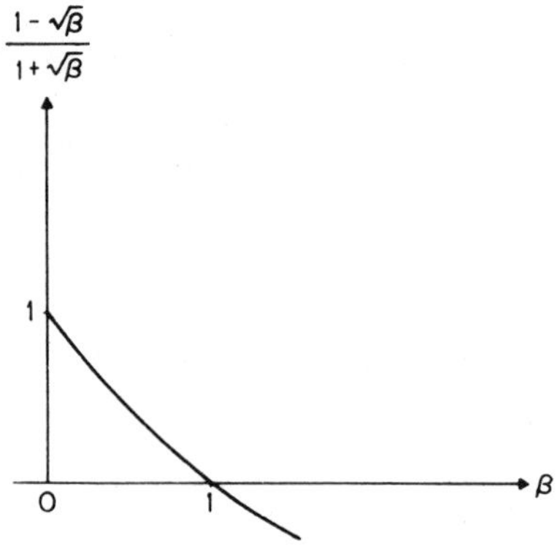

FIG. 5.2.2.b. Plot of α as a function of β.

5.2.3. DUALITY OF THE SOLUTION FOR $\epsilon \neq 1$

When the parameter ϵ is not equal to one, the preceding proof is not valid, but the boundary value problem can be solved numerically. In this connection, we consider the equations

$$\begin{aligned} F''' + FF'' - GG'' &= 0 \\ G'' + \epsilon(FG' - F'G) &= 0 \end{aligned} \tag{5.2.3-1}$$

and the boundary conditions

$$F(0) = F'(0) = G(0) = 0, \qquad F''(0) = p, \qquad G'(0) = 1$$

The parameter p being chosen arbitrarily, we integrate Eq. (5.2.3-1) numerically, using the preceding boundary conditions. As a result, we get the values of $F'(\infty)$ and $G'(\infty)$; let $F'(\infty) = 2A$, $G'(\infty) = 2B$.

The functions

$$f(\eta) = A^{-1/2}F(A^{-1/2}\eta)$$
$$g(\eta) = (A^{1/2}B^{-1})G(A^{-1/2}\eta)$$

satisfy the system (5.2.1-5) with $\beta = B^2/A^2$ as well as the boundary conditions

$$f(0) = f'(0) = 0, \qquad g(0) = 0$$
$$f'(\infty) = 2, \qquad g'(\infty) = 2$$

We find the value of β at the end of the calculation, and the problem consists in choosing p so that the final value of β agrees with that given in advance. Both A and B, and hence β, are functions of p. The function $f''(0) = pA^{-3/2}$ is also a function of p. When ϵ is fixed, we get a solution for each value of p. The corresponding values of β and $f''(0)$ are determined at the end of the computation. Figure 5.2.3.a shows

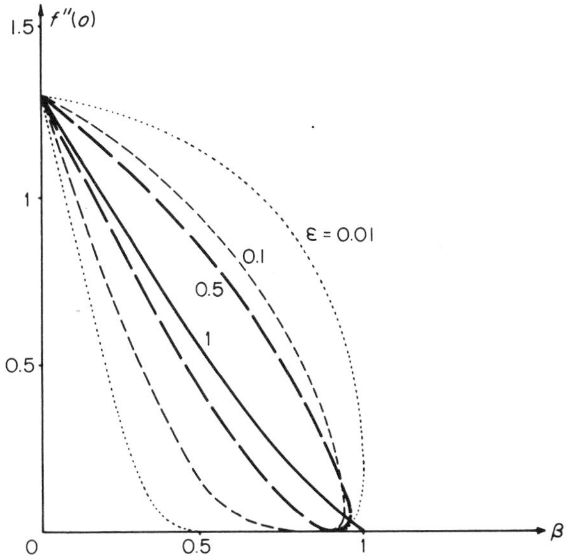

FIG. 5.2.3.a. Locus of points with coordinates β and $f''(0)$.

the locus of the points with coordinates β and $f''(0)$ for different values of ϵ, when p varies.

For $\epsilon = 1$, only one value of $f''(0)$, hence, one value of p, corresponds to a given value of β between 0 and 1. We recover the existence and uniqueness theorem proved in the previous section.

For $0 < \epsilon < 1$, we find two values of $f''(0)$, hence, two values of p, corresponding to each value of β between 0 and $\beta_{\max}$. The maximum values of β are shown in Fig. 5.2.3.b. For $0 < \beta < \beta_{\max}$, the boundary value problem admits two solutions; the first one corresponding to the large value of $f''(0)$ and the second one to the small value of $f''(0)$. For these two solutions, we show the variation of the component of velocity u (Fig. 5.2.3.c) and of the magnetic field H_x (Fig. 5.2.3d) parallel to the plate. The values U and H correspond to infinity.

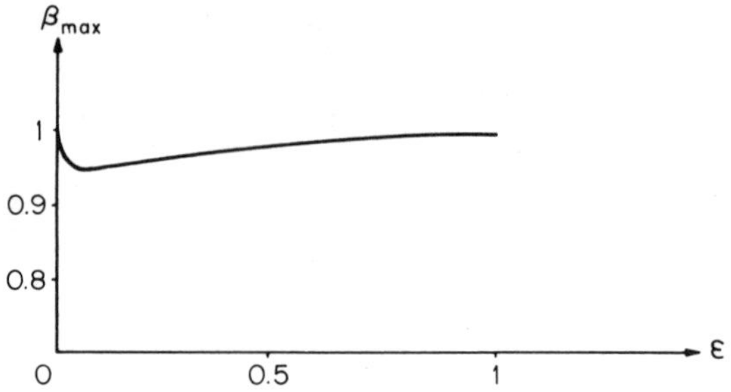

FIG. 5.2.3.b. Maximum values of β.

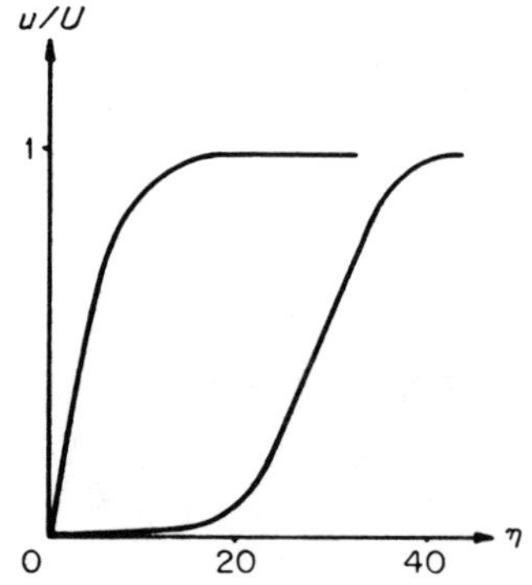

FIG. 5.2.3.c. Left curve: first solution; right curve: second solution. $\beta = 0.8$, $\epsilon = 0.1$.

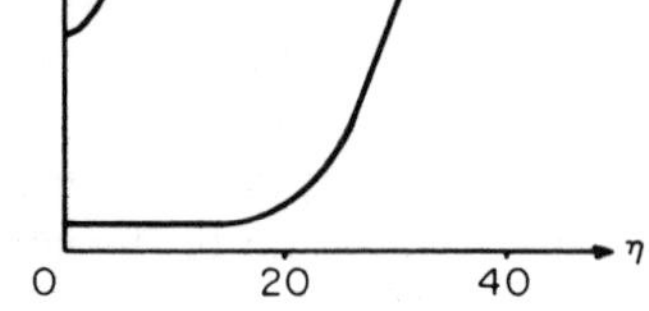

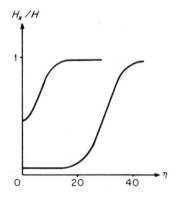

FIG. 5.2.3.d. Left curve: first solution; right curve: second solution. $\beta = 0.8$, $\epsilon = 0.1$.

The second solution corresponds to a thicker boundary layer; for this solution, the drag is lower and the electric current larger.

5.3. Flow Past a Flat Plate of Finite Length

5.3.1. Integral Equation of the Problem

The equations of motion for plane flow of an incompressible fluid can be written in the form (5.2.1-3); these equations admit the solution $\Psi = 1$, $A = 1$, which corresponds to a uniform flow, and the change of the function

$$\Psi = y + \delta\Psi, \qquad A = y + \delta A$$

leads to the following linear equations in $\delta\Psi$ and δA (δ is omitted for simplicity)

$$\begin{aligned} \Delta\Delta\Psi - \Delta(\Psi_t + \Psi_x - \beta A_x) &= 0 \\ \Delta A - \epsilon(A_t + A_x - \Psi_x) &= 0 \end{aligned} \tag{5.3.1-1}$$

The components of the fluid velocity and the magnetic field are

$$\mathbf{H} \left| \begin{aligned} H_x &= 1 + A_y \\ H_y &= -A_x \\ H_z &= 0 \end{aligned} \right. , \qquad \mathbf{V} \left| \begin{aligned} u &= 1 + \Psi_y \\ v &= -\Psi_x \\ w &= 0 \end{aligned} \right.$$

We propose to study those solutions of Eqs. (5.3.1-1) which represent steady flow past a finite plate of length l, assumed to be along the axis of the abscissa. In a dimensionless coordinate system, the plate is represented by the equations

$$y = 0, \qquad 0 \leqslant x \leqslant R = V_0 l/\nu \quad \text{(Reynolds number)}$$

The velocity field and the acceleration field are assumed to be continuous throughout the fluid, i.e., throughout the plane, except, in the limit, at the flat plate. The condition of no slip at the plate gives:

$$\frac{\partial\Psi}{\partial x}(x, 0) = 0, \qquad \frac{\partial\Psi}{\partial y}(x, 0) = -1, \qquad 0 \leqslant x \leqslant R$$

It follows that

$$\frac{\partial^2\Psi}{\partial x^2}(x, 0) = \frac{\partial^2\Psi}{\partial x\,\partial y}(x, 0) = 0, \qquad 0 \leqslant x \leqslant R$$

The four derivatives Ψ_x, Ψ_y, Ψ_{xx}, and Ψ_{xy} are continuous throughout the plane, while the derivative Ψ_{yy} can eventually be discontinuous at the plate. We assume that

$$f(x) = 0, \qquad x(x - R) > 0$$

$$f(x) = \left[\frac{\partial^2 \Psi}{\partial y^2}(x, 0)\right], \qquad 0 \leqslant x \leqslant R$$

The symbol $[Q]$ denotes the difference $Q(x, 0^+) - Q(x, 0^-)$. The condition $(\partial\Psi/\partial y)(x, 0) = -1$ may seen to contradict the linearization of the equations. This is not true if the motion is slow, with a small value of the free-stream velocity, for then the condition of no slip can be satisfied with a velocity that stays small. We will therefore assume that the motion is slow.

At infinity, the velocity and magnetic field are uniform

$$\frac{\partial A}{\partial x} = \frac{\partial A}{\partial y} = 0, \qquad \frac{\partial \Psi}{\partial x} = \frac{\partial \Psi}{\partial y} = 0$$

In the case of steady flow, Eqs. (5.3.1-1) become

$$\begin{aligned} \Delta\Delta\Psi - \Delta(\Psi_x - \beta A_x) &= 0 \\ \Delta A - \epsilon(A_x - \Psi_x) &= 0 \end{aligned} \tag{5.3.1-2}$$

Taking the Laplace transform of these equations, we get algebraic relations. If we write

$$\begin{aligned} \bar{Q}_1(r) &= \int_0^\infty Q_1(y)\, e^{-iry}\, dy \\ &= \left| \frac{Q_1(y)\, e^{-iry}}{-ir} \right|_0^\infty + \int_0^\infty \frac{\partial Q_1}{\partial y} \frac{e^{-iry}}{i\pi}\, dy \end{aligned}$$

we have

$$\overline{\partial Q_1/\partial y} = ir\bar{Q}_1 - Q_1(0^+)$$

provided $Q_1(\infty)$ is zero. Similarly, if we write

$$\bar{Q}_2(r) = \int_{-\infty}^0 Q_2(y)\, e^{-iry}\, dy$$

we get, if $Q_2(\infty) = 0$,

$$\overline{\partial Q_2/\partial y} = ir\bar{Q}_2 + Q_2(0^-)$$

If we write now

$$\bar{Q}(p, r) = \iint_{-\infty}^{\infty} Q(x, y) \exp\{-i(px + ry)\}\, dx\, dy$$

and if $Q(x, y)$ becomes zero at infinity and is continuous throughout the plane except possibly on the line $y = 0$, we have

$$\overline{\partial Q/\partial y} = ir\bar{Q} - \int_{-\infty}^{\infty} [Q(x, 0)]\, e^{-ipx}\, dx$$

At infinity, the function $\Psi(x, y)$ is not zero, but its derivative, $\partial\Psi(x, y)/\partial x$, is zero. We can define $\overline{\partial\Psi/\partial x}$, which we denote by $ip\bar{\Psi}$ (which defines $\bar{\Psi}$):

$$\overline{\partial\Psi/\partial x} = ip\bar{\Psi}$$

Since

$$\frac{\partial}{\partial y}\left(\frac{\partial\Psi}{\partial x}\right) = \frac{\partial}{\partial x}\left(\frac{\partial\Psi}{\partial y}\right)$$

we can write $ir(ip\bar{\Psi}) = ip\ \overline{\partial\Psi/\partial y}$; hence, $\overline{\partial\Psi/\partial y} = ir\bar{\Psi}$. We then find

$$\overline{\partial^2\Psi/\partial x^2} = -p^2\bar{\Psi}, \qquad \overline{\partial^2\Psi/\partial y^2} = -r^2\bar{\Psi}$$

$$\overline{\partial(\Delta\Psi)/\partial x} = -ip(p^2 + r^2)\,\bar{\Psi}$$

$$\overline{\partial(\Delta\Psi)/\partial y} = -ir(p^2 + r^2)\,\bar{\Psi} - \bar{f}(p)$$

where

$$\bar{f}(p) = \int_{-\infty}^{\infty} f(x)\, e^{-ipx}\, dx = \int_{0}^{R} f(x)\, e^{-ipx}\, dx$$

$$\overline{\Delta\Delta\Psi} = (p^2 + r^2)^2\,\bar{\Psi} - ir\bar{f}(p)$$

Applying the Laplace transform in two variables to Eqs. (5.3.1-2), we get

$$\begin{aligned}(p^2 + r^2)\{(p^2 + r^2 + ip)\,\bar{\Psi} - ip\beta\bar{A}\} &= ir\bar{f}\\ (p^2 + r^2 + ip\epsilon)\,\bar{A} - ip\epsilon\bar{\Psi} &= 0\end{aligned} \tag{5.3.1-3}$$

It then follows that

$$\begin{aligned}\bar{\Psi} &= \frac{ir\bar{f}}{p^2 + r^2}\,\frac{p^2 + r^2 + i\epsilon p}{(p^2 + r^2 + ip)(p^2 + r^2 + i\epsilon p) + \epsilon\beta p^2}\\ \bar{A} &= \frac{-\epsilon pr\bar{f}}{p^2 + r^2}\,\frac{1}{(p^2 + r^2 + ip)(p^2 + r^2 + i\epsilon p) + \epsilon\beta p^2}\end{aligned} \tag{5.3.1-4}$$

Since the function $\overline{\Psi}(p, r)$ is known, we have $\overline{(\partial\Psi/\partial y)} = ir\overline{\Psi}$, and taking the inverse Laplace transform,

$$\frac{\partial\Psi}{\partial y}(x, y) = \iint e^{i(px+ry)} ir\overline{\Psi}(p, r)\, dp\, dr$$

$$\frac{\partial\Psi}{\partial y}(x, y) = \int e^{ipx}\overline{K}(p, y)\bar{f}(p)\, dp$$

with

$$\overline{K}(p) = \int e^{iry} ir \frac{ir}{p^2 + r^2} \frac{(p^2 + r^2 + i\epsilon p)\, dr}{(p^2 + r^2 + ip)(p^2 + r^2 + i\epsilon p) + \epsilon\beta p^2}$$

The product $\overline{K}(p)\bar{f}(p)$ is the Laplace transform of the function $\partial\Psi(x, y)/\partial y$. By the convolution theorem, we can write

$$\frac{\partial\Psi}{\partial y}(x, y) = \int_0^R K(x - \xi, y; \epsilon; \beta) f(\xi)\, d\xi$$

where

$$\begin{aligned} \overline{K}(x, y; \epsilon, \beta) &= \int e^{ipx} K(p)\, dp \\ &= \iint e^{i(px+ry)} \frac{-r^2}{p^2 + r^2} \frac{p^2 + r^2 + i\epsilon p}{(p^2 + r^2 + ip)(p^2 + r^2 + i\epsilon p) + \epsilon\beta p^2}\, dp\, dr \end{aligned}$$

To take account of boundary conditions on the plate, it is sufficient to state that the derivative $\partial\Psi(x, y)/\partial x$ equals -1 when $y = 0$ and $0 \leqslant x \leqslant R$. This condition leads to the integral equation

$$\int_0^R K(x - \xi, 0; \epsilon, \beta) f(\xi)\, d\xi = -1 \tag{5.3.1-5}$$

Once this integral equation is solved, the function $f(x)$ is known and the calculation can be completed without further theoretical difficulties.

The function $K(x, y; \epsilon, \beta)$ can be expressed in terms of modified Bessel functions of the second kind K_0 and K_1. We have

$$K_0(z) = \int_1^\infty e^{-zt} \frac{dt}{(t^2 - 1)^{1/2}}$$

$$K_1(z) = -\frac{dK_0}{dz} = z \int_1^\infty e^{-zt} (t^2 - 1)^{1/2}\, dt$$

The behavior of the functions K_0 and K_1 is shown in Fig. 5.3.1.a. We then write

$$H(x, y) = -\frac{x}{2\pi(x^2+y^2)} + \frac{1}{4\pi} e^{x/2} \times \left\{K_0\left(\frac{(x^2+y^2)^{1/2}}{2}\right) + \frac{x}{(x^2+y^2)^{1/2}} K_1\left[\frac{(x^2+y^2)^{1/2}}{2}\right]\right\}$$

$$\lambda = \{(1-\epsilon^2)^2 + 4\,\epsilon\,\beta\}^{1/2}$$

$$2\,\Omega^2 = \lambda - 1 - \epsilon, \qquad 2\,\omega^2 = \lambda + 1 + \epsilon$$

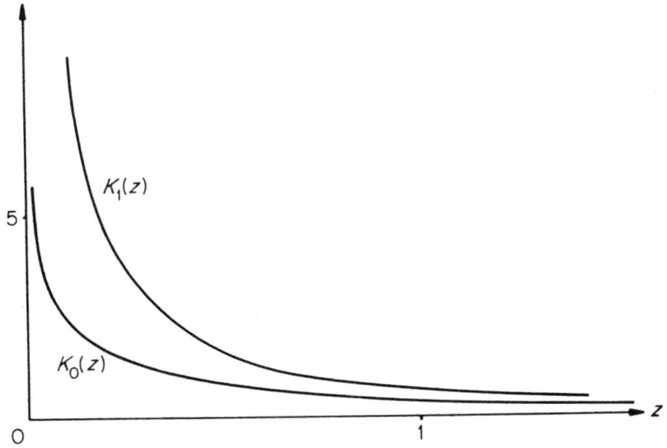

FIG. 5.3.1.a. Functions $K_0(z)$ and $K_1(z)$.

We then find

$$K(x, y; \epsilon, \beta) = \tfrac{1}{2}(\lambda + \epsilon - 1)\, H(-\Omega^2 x, y) + \tfrac{1}{2}(\lambda - \epsilon - 1)\, H(\omega^2 x, y)$$

and hence

$$K(x, 0; \epsilon, \beta) = -\frac{1}{2\pi x} + \frac{1}{4\pi} e^{x/2} \left\{K_0\left(\frac{|x|}{2}\right) + \frac{x}{|x|} K_1\left(\frac{|x|}{2}\right)\right\}$$

5.3.2. The Case of Large Reynolds Numbers

In the general case, the integral equations (5.3.1-5) can be solved only by numerical methods. We can, however, find analytical solutions when the Reynolds number is either very large or very small.

If we put $x = Rz$ and $\xi = R\zeta$, the integral equation (5.3.1-5) can be written

$$\int_0^1 K\{R(z-\zeta);\, \epsilon, \beta\} f(R\zeta)\, R\, d\zeta = 1 \tag{5.3.2-1}$$

where $0 \leqslant z \leqslant 1$. When the product $R(z-\zeta)$ is large, i.e., when R is large and $(z-\zeta)$ is different from zero, the expression for the kernel K can be simplified if we replace the function $H(x, 0)$ by the first terms of its asymptotic expansion.

For large values of z, we have†

$$K_\mu(z) \sim \left(\frac{\pi}{2z}\right)^{1/2} e^{-z} \left\{1 + \frac{4\nu^2 - 1}{1!\, 8z} + \frac{(4\nu^2-1)(4\nu^2-3)}{2!\,(8z)^2} + \cdots\right\}$$

We deduce that

$$H(x, 0) \sim -\frac{1}{2\pi x} + O\left(\frac{e^{-|x|}}{(|x|)^{1/2}}\right) \qquad \text{for} \quad x \ll -1$$

$$H(x, 0) \sim \frac{1}{2(\pi x)^{1/2}} - \frac{1}{2\pi x} + O(x^{-3/2}) \qquad \text{for} \quad x \gg 1$$

or

$$K\{R(z-\zeta), 0;\, \epsilon, \beta\} \sim \frac{\lambda - \epsilon + 1}{4\lambda\omega} \frac{1}{[R(z-\zeta)]^{1/2}} \qquad \text{for} \quad 0 \leqslant \zeta < z$$

$$K\{R(z-\zeta), 0;\, \epsilon, \beta\} \sim \frac{\lambda + \epsilon + 1}{4\lambda\Omega} \frac{1}{[R(\zeta - z)]^{1/2}} \qquad \text{for} \quad z < \zeta \leqslant 1$$

When the value of ζ is very near that of z, the preceding expansions are not valid, but we can verify that the corresponding values of ζ make negligible contributions to the integral. Hence, for large values of the Reynolds number, Eq. (5.3.2-1) can be replaced by the following simplified equation:

$$\int_0^z \frac{g(\zeta)\, d\zeta}{(z-\zeta)^{1/2}} + \alpha \int_z^1 \frac{g(\zeta)\, d\zeta}{(\zeta - z)^{1/2}} = 1 \tag{5.3.2-2}$$

where

$$g(\zeta) = \frac{\lambda - \epsilon + 1}{4\lambda\omega} \left(\frac{R}{\pi}\right)^{1/2} f(R\zeta)$$

$$\alpha = \frac{\lambda + \epsilon - 1}{\lambda - \epsilon + 1} \frac{\omega}{\Omega}$$

† Watson, G., "A Treatise on the Theory of Bessel Functions," Cambridge (1944), p. 202, formula (1).

Equation (5.3.2-2) can be further simplified by the method of Abel, which consists of multiplying both sides by $dz/(x - z)^{1/2}$ and integrating from 0 to x. This will yield

$$\int_0^x \frac{dz}{(x-z)^{1/2}} \int_0^z \frac{g(\zeta)\, d\zeta}{(z-\zeta)^{1/2}} + \alpha \int_0^x \frac{dz}{(x-z)^{1/2}} \int_z^1 \frac{g(\zeta)\, d\zeta}{(\zeta - z)^{1/2}} = 2\sqrt{x} \tag{5.3.2-3}$$

The first term can be transformed by changing the order of integration; working in the ζ, z plane (Fig. 5.3.2.a), we get

$$\int_0^x \frac{dz}{(x-z)^{1/2}} \int_0^z \frac{g(\zeta)\, d\zeta}{(z-\zeta)^{1/2}} = \int_0^x g(\zeta)\, d\zeta \int_\zeta^x \frac{dz}{[(x-z)(z-\zeta)]^{1/2}}$$

$$= \pi \int_0^x g(\zeta)\, d\zeta$$

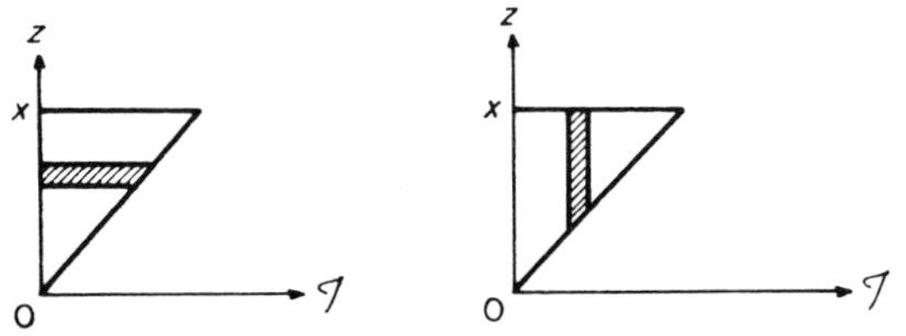

FIG. 5.3.2.a. Transformation of first term, Eq. (5.3.2-3).

The second term will also simplify by the same procedure; working in the ζ, z plane (Fig. 5.3.2.b), we get

$$\int_0^x \frac{dz}{(x-z)^{1/2}} \int_z^1 \frac{g(\zeta)\, d\zeta}{(\zeta - z)^{1/2}} = \int_0^x g(\zeta)\, d\zeta \int_0^\zeta \frac{dz}{[(x-z)(\zeta - z)]^{1/2}}$$

$$+ \int_x^1 g(\zeta)\, d\zeta \int_0^x \frac{dz}{[(x-z)(\zeta-z)]^{1/2}}$$

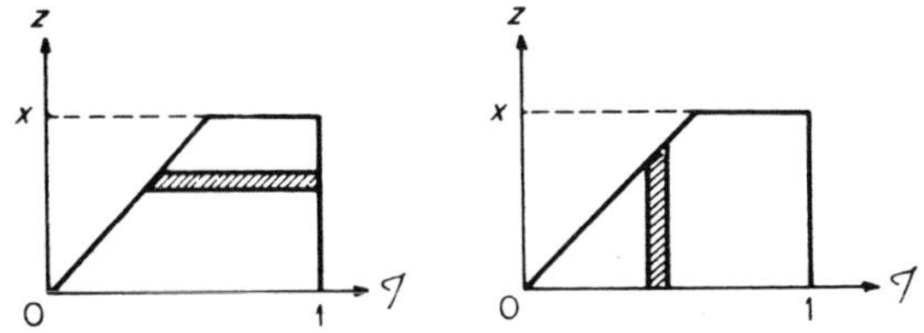

FIG. 5.3.2.b. Transformation of second term, Eq. (5.3.2-3).

or

$$\int_0^x \frac{dz}{(x-z)^{1/2}} \int_z^1 \frac{g(\zeta)\, d\zeta}{(\zeta - z)^{1/2}} = \int_0^1 g(\zeta) \operatorname{argch} \frac{x+\zeta}{|x-\zeta|}\, d\zeta$$

Differentiating both sides of Eq. (5.3.2-3) with respect to x yields

$$\pi g(x) = \frac{1}{\sqrt{x}} - \alpha \int_0^1 \left(\frac{\zeta}{x}\right)^{1/2} \frac{g(\zeta)}{\zeta - x}\, d\zeta$$

Since x lies between 0 and 1, the integral on the right-hand side can be written

$$\lim_{\epsilon \to 0} \left\{ \int_0^{x-\epsilon} \left(\frac{\zeta}{x}\right)^{1/2} \frac{g(\zeta)}{\zeta - x}\, d\zeta + \int_{x+\epsilon}^1 \left(\frac{\zeta}{x}\right)^{1/2} \frac{g(\zeta)}{\zeta - x}\, d\zeta \right\}$$

The change of function $h(x) = x^{1/2} g(x)$ yields the simpler form

$$\pi h(x) = 1 - \alpha \int_0^1 \frac{h(\zeta)}{\zeta - x}\, d\zeta \tag{5.3.2-4}$$

Equation (5.3.2-4) is of the general type of integral equations studied by Carleman. According to Carleman's theory, Eq. (5.3.2-4) possesses one and only one solution with an integrable singularity; to obtain this solution, we compute the integral

$$\int_0^1 \frac{h^*(\zeta)}{\zeta - x}\, d\zeta \qquad \text{with} \qquad h^*(z) = \left(\frac{z}{1-z}\right)^r, \quad 0 < r < 1$$

For this, we will consider, in the complex plane $\zeta = \xi + i\eta$, the contour Γ formed by the segments ϵ, $x - \epsilon$, $x + \epsilon$, and $1 - \epsilon$ of the real axis (traversed twice along each segment) and the circles of radius ϵ with centers respectively at 0, x, and 1 (Fig. 5.3.2.c). When

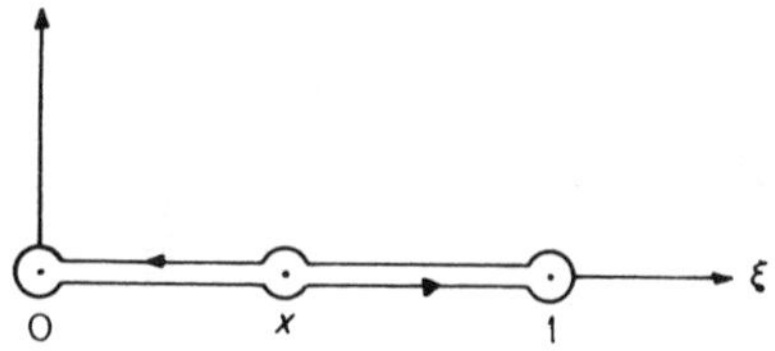

FIG. 5.3.2.c. The contour Γ in the complex plane $\zeta = \xi + i\eta$.

the point corresponding to ζ moves in the positive direction once around Γ, the function

$$\frac{h^*(\zeta)}{\zeta - x} = \left(\frac{\xi}{1-\zeta}\right)^r \frac{1}{\zeta - x}$$

which is many-valued, is multiplied by $e^{2i\pi r}$ when rotating about the origin and by $e^{-2i\pi r}$ when rotating about the point $\zeta = 1$. Hence, it takes its initial value after one circuit. The integral along the upper semicircle, center x, has the value $i\pi R_x$,

$$R_x = \left(\frac{x}{1-x}\right)^r$$

being the residue at the point $\zeta = x$; the integral along the lower semicircle, center x, takes the value $e^{2i\pi r} i\pi R_x$. The integral around the circles with centers $\zeta = 0$ and $\zeta = 1$ is equal to zero. The integral taken around Γ is equal to that taken in the positive direction along a circle centered at the origin with very large radius R; writing $\zeta = Re^{i\theta}$, we have

$$\begin{aligned}\int \frac{h(\zeta)}{\zeta - x}\, d\zeta &= \int_0^{2\pi} \left(\frac{Re^{i\theta}}{1 - Re^{i\theta}}\right)^r \frac{Re^{i\theta}}{Re^{i\theta} - x}\, i\, d\theta \\ &= e^{i\pi r} \int_0^{2\pi} \left(1 - \frac{1}{Re^{i\theta}}\right)^{-r} \left(1 - \frac{x}{Re^{i\theta}}\right)^{-1} i\, d\theta \\ &= e^{i\pi r} \int_0^{2\pi} \left(1 + \frac{r+x}{Re^{i\theta}} + \cdots\right) i\, d\theta = 2i\pi e^{i\pi r}\end{aligned}$$

and we get

$$(e^{2i\pi r} - 1) \int_0^1 \left(\frac{\zeta}{1-\zeta}\right)^r \frac{d\zeta}{\zeta - x} + (e^{2i\pi r} + 1)\, i\pi \left(\frac{x}{1-x}\right)^r = 2i\pi e^{i\pi r}$$

or

$$h^*(x) \cos \pi r = 1 - \frac{\sin \pi r}{\pi} \int_0^1 \frac{h^*(\zeta)}{\zeta - x}\, d\zeta$$

Hence, the function

$$h(x) = \frac{\cos \pi r}{\pi} h^*(x) = \frac{\cos \pi r}{\pi} \left(\frac{x}{1-x}\right)^r$$

will be the solution (with integrable singularities) of the integral equation (5.3.1-4) if we choose r such that $\tan(\pi r) = \alpha$. We can then deduce the expressions for the functions $g(x)$ and $f(x)$,

$$\begin{aligned} g(x) &= \frac{\cos \pi r}{\pi} \frac{1}{(\pi x)^{1/2}} \left(\frac{x}{1-x}\right)^r \\ f(x) &= \frac{4\lambda\omega}{\lambda + 1 - \epsilon} \frac{\cos \pi r}{(\pi x)^{1/2}} \left(\frac{x}{R-x}\right)^r \end{aligned} \tag{5.3.2-5}$$

We have, in particular:

For a perfect insulator $(\epsilon = 0)$, $f(x) = 2/(\pi x)^{1/2}$.

For a perfect conductor $(\epsilon = \infty)$, $f(x) = [(\beta - 1)/\pi(R - x)]^{1/2}$.

5.3.3. The Case of Small Reynolds Numbers

When the Reynolds number R is very small, the kernel of the integral Eq. (5.3.1-5) can be simplified, because the variable ξ varies between 0 and R, and x must also have a value between 0 and R. For small values of x,

$$H(x, 0) = -\frac{1 - \gamma + \log(|x|/4)}{4\pi} + O(x \log |x|)$$

where γ is Euler's constant. The kernel $K(x - \xi, 0, \epsilon, \beta)$ then becomes a linear function of $\log |x - \xi|$ and the integral Eq. (5.3.1-5) takes the simplified form

$$\int_0^R A(x - \xi) f(\xi)\, d\xi = -8\pi\lambda \tag{5.3.3-1}$$

where

$$\begin{aligned} A(x - \xi) = (\lambda + \epsilon - 1) &\left\{1 - \gamma + \log \frac{\Omega^2}{4} |x - \xi|\right\} \\ + (\lambda - \epsilon + 1) &\left\{1 - \gamma + \log \frac{\omega^2}{4} |x - \xi|\right\} \end{aligned}$$

We again find an integral equation of Carleman's type, the solution of which has been shown to exist and to be unique; we will find this solution.

We again consider in the complex plane $\zeta = \xi + i\eta$ the contour Γ of Fig. 5.3.2.c, and the integral along this contour of the function

$$\log\left(1 - \frac{x}{\zeta}\right) \frac{1}{[\zeta(1 - \zeta)]^{1/2}}$$

When we make a half turn in the positive sense about the point $z = x$, the function $\log(\zeta - x)$ becomes $\log(\zeta - x) + i\pi$. When we rotate about the origin, the function $\log \zeta$ becomes $\log \zeta + 2i\pi$, and the function $\zeta^{-1/2}$ becomes $\zeta^{-1/2}e^{-i\pi}$; when we turn about the point $\zeta = 1$, the function $(1 - \zeta)^{-1/2}$ becomes $(1 - \zeta)^{-1/2}e^{-i\pi}$. The integrals on the three circles of radius ϵ are zero. We then have

$$\int_{(\Gamma)} \log\left(1 - \frac{x}{\zeta}\right) \frac{d\zeta}{[\zeta(1 - \zeta)]^{1/2}}$$

$$= \int_1^x \frac{\log|\xi - x| - \log \xi}{[\xi(1 - \xi)]^{1/2}} d\xi$$

$$+ \int_x^0 \frac{\log|\xi - x| + i\pi - \log \xi}{[\xi(1 - \xi)]^{1/2}} d\xi$$

$$- \int_0^x \frac{\log|\xi - x| + i\pi - \log \xi - 2i\pi}{[\xi(1 - \xi)]^{1/2}} d\xi$$

$$- \int_x^1 \frac{\log|\xi - x| + i\pi + i\pi - \log \xi - 2i\pi}{[\xi(1 - \xi)]^{1/2}} d\xi$$

$$= -2 \int_0^1 \frac{\log|\xi - x| - \log \xi}{[\xi(1 - \xi)]^{1/2}} d\xi$$

But the integral on the contour Γ is equal to that along a circle of large radius with center at the origin. Since this integral is zero, we get

$$\int_0^1 \frac{\log|\xi - x|}{(1 - \xi)} d\xi = \int_0^1 \frac{\log \xi}{\xi(1 - \xi)} d\xi$$

The value I of the last integral can be obtained by writing $\xi = \sin^2 \phi$ or $\xi = \cos^2 \phi$; we get

$$I = 4 \int_0^{\pi/2} \log \sin \varphi \, d\varphi = 4 \int_0^{\pi/2} \log \cos \varphi \, d\varphi$$

$$2I = 4 \int_0^{\pi/2} \log(\sin \varphi \cos \varphi) \, d\varphi = 2 \int_0^{\pi} (\log \sin \Psi - \log 2) \, d\Psi$$

$$2I = 4 \int_0^{\pi/2} \log \sin \Psi \, d\Psi - 2\pi \log 2 = I - 2\pi \log 2$$

We deduce that

$$\int_0^1 \frac{\log|\xi - x|}{[\xi(1 - \xi)]^{1/2}} d\xi = -2\pi \log 2 \qquad (5.3.3\text{-}2)$$

which enables us to solve Eq. (5.3.3-1). The solution will have the form

$$f(x) = -\frac{4}{\Lambda}\frac{1}{[x(R-x)]^{1/2}} \tag{5.3.3-3}$$

where the constant Λ is given by the relation

$$\Lambda = \gamma - 1 + \log\frac{R}{16} + \frac{\lambda+\epsilon-1}{\lambda}\log\Omega + \frac{\lambda-\epsilon+1}{\lambda}\log\omega$$

5.3.4. Calculation of the Drag

To compute the forces on the plate, we will determine first the stress tensor $\mathbf{T}$; we then revert to dimensional variables, letting x, y stay dimensionless:

$$\mathbf{T} = -p\mathbf{U} + 2\mu_1\mathbf{D}, \qquad \mu_1 = \nu\rho_0$$

$$2\mathbf{D}_1 = \nabla\mathbf{V} + \overline{\nabla\mathbf{V}} = (V_0^2/\nu)\begin{vmatrix} 2\Psi_{xy} & \Psi_{yy}-\Psi_{xx} \\ \Psi_{yy}-\Psi_{xx} & -2\Psi_{xy}\end{vmatrix}$$

The pressure $\mathbf{p}$ on the plate is equal to the product $\mathbf{T}\cdot\mathbf{n}$, which gives

$$\text{for } y = 0^+, \qquad \mathbf{p} = \begin{vmatrix} \rho_0 V_0^2(\Psi_{yy}-\Psi_{xx}) \\ -p - 2\rho_0 V_0^2\,\Psi_{xy}\end{vmatrix}$$

$$\text{for } y = 0^-, \qquad \mathbf{p} = \begin{vmatrix} -\rho_0 V_0^2(\Psi_{yy}-\Psi_{xx}) \\ p + 2\rho_0 V_0^2\,\Psi_{xy}\end{vmatrix}$$

If we then add these equations, we obtain

$$p_x = \rho_0 V_0^2[\Psi_{yy}], \qquad p_y = 0$$

Only the component of the pressure parallel to the plate is different from zero. This component defines the friction coefficient

$$p_x = \rho_0 V_0^2[\Psi_{yy}] = \rho_0 V_0^2 f(x)$$

The drag T and the drag coefficient C_x are then defined by the formula

$$T = \rho_0 V_0 \nu \int_0^R f(x)\,dx \tag{5.3.4-1}$$

$$C_x = \frac{T}{\rho_0 V_0^2 l} = \frac{1}{R}\int_0^R f(x)\,dx \tag{5.3.4-2}$$

It is interesting to write the results explicitly, replacing $f(x)$ by the approximate values obtained when the Reynolds number is either very large or very small.

a. The Case of Large Reynolds Number

For this case,

$$f(x) = \frac{4\lambda\omega}{\lambda + 1 - \epsilon} \frac{\cos \pi r}{\sqrt{\pi x}} \left(\frac{x}{R - x}\right)^r$$

The singularity is $x^{r-(1/2)}$ at the leading edge, $(R - x)^{-r}$ at the trailing edge. The second singularity is more important for $r < 1/4$ or $\alpha < 1$.

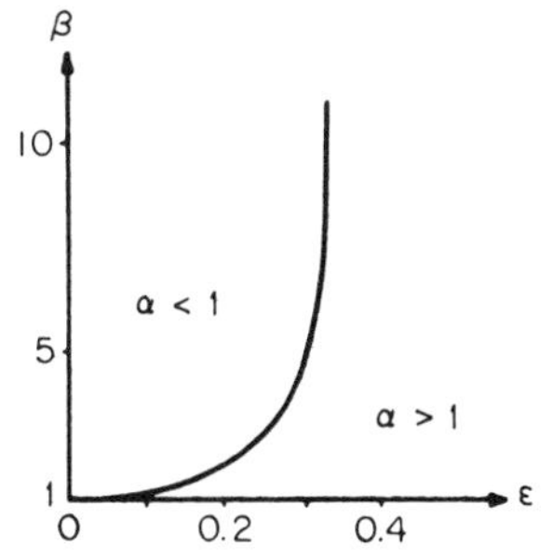

FIG. 5.3.4.a. The region $\alpha < 1$: Predominant downstream wake; $\alpha > 1$: Predominant upstream wake.

In the plane of the variables ϵ, β (Fig. 5.3.4.a), the condition $\alpha = 1$ is written

$$4\beta (3\epsilon - 1) + (\epsilon - 1)^2 (\epsilon + 1)(3\epsilon + 2) = 0$$

and determines two regions. In the region $\alpha < 1$, where the importance of the trailing-edge singularity is more pronounced, the downstream wake will be more important than the upstream wake; the opposite is true for $\alpha > 1$. For $r = 0$ let $\epsilon = 0$ (the case of perfect insulator), the upstream wake disappears; for $r = 1/2$ let $\epsilon = \infty$ (perfect conductor), it is the downstream wake which disappears, while only the upstream wake exists.

The variation along the plate of the coefficient of friction is shown in Fig. 5.3.4.b for different values of the electric conductivity (parameter ϵ) and for a fixed value of magnetic field (parameter β).

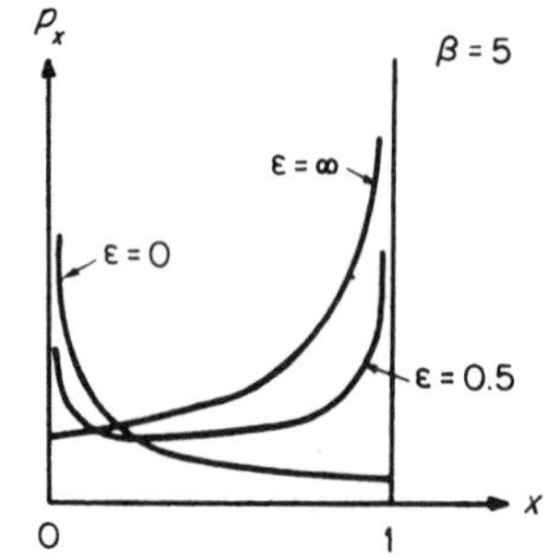

FIG. 5.3.4.b. Variation of the coefficient of friction.

The drag coefficient C_x can be written in terms of the gamma function; we have

$$\int_0^1 x^{r-(1/2)}(1-x)^{-r}\,dx = 2\pi^{-1/2}\Gamma(\tfrac{1}{2}+r)\,\Gamma(1-r)$$

and hence

$$C_x = \frac{8}{\pi(\lambda+1-\epsilon)}\,\frac{1}{[R(1+\alpha^2)]^{1/2}}\,\Gamma(\tfrac{1}{2}+r)\,\Gamma(1-r) \qquad (5.3.4\text{-}3)$$

In particular, for a perfect insulator ($\epsilon = 0$), $C_x = 4/(\pi R)^{1/2}$; for a perfect conductor ($\epsilon = \infty$), $C_x = 4[(\beta - 1)/\pi R]^{1/2}$

b. The Case of Small Reynolds Numbers

For this case,

$$f(x) = -\frac{4}{\Lambda}\,\frac{1}{[x(R-x)]^{1/2}}, \qquad C_x = \frac{-4\pi}{\Lambda R} \qquad (5.3.4\text{-}4)$$

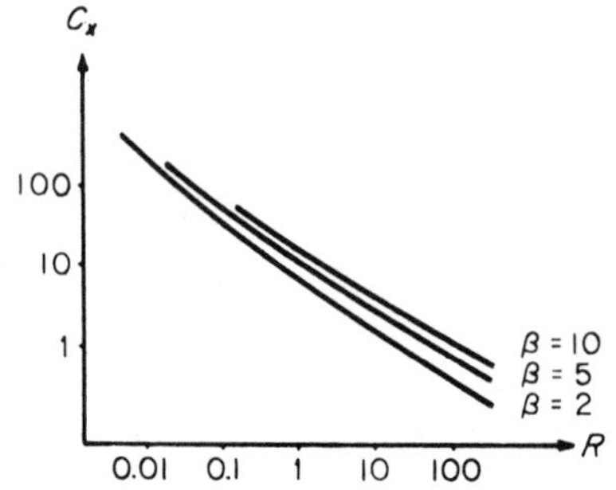

FIG. 5.3.4.c. Perfect conductor, $\epsilon = \infty$.

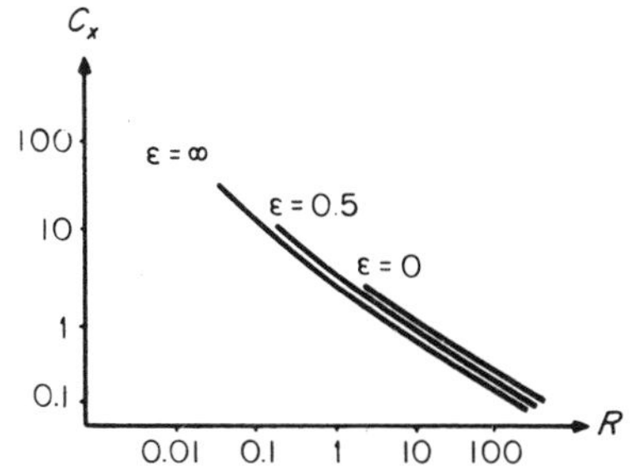

FIG. 5.3.4.d. Weak magnetic field, $\beta = 0.5$.

In Fig. 5.3.4.c, the variation of the drag coefficient as a function of the Reynolds number is shown for a perfect conductor; the flow is always sub-Alfvénic, $\beta > 1$. In Figs. 5.3.4.d and 5.3.4.e, the variations of the drag coefficient as a function of Reynolds numbers are shown for a given magnetic field.

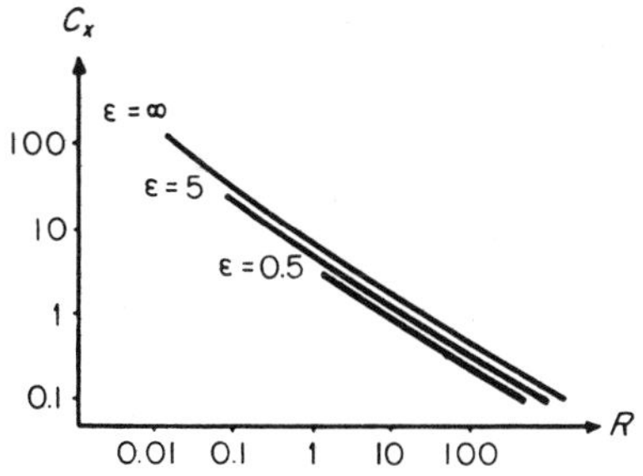

FIG. 5.3.4.e. Strong magnetic field, $\beta = 5$.

5.4. Flow Past a Magnetized Sphere

5.4.1. Equations of the Problem

An invisicid incompressible fluid of density ρ_0 and electrical conductivity σ moves with a uniform velocity $\mathbf{V}_\infty$ at infinity. A magnetized sphere of radius a is placed in this fluid; the sphere is a dipole of magnetic moment M.

The flow being steady, the equations of motion reduce to

$$(\nabla \times \mathbf{V}) \times \mathbf{V} + \nabla \left(\frac{V^2}{2} + \frac{p}{\rho_0} \right) = \frac{\mu}{\rho_0} (\nabla \times \mathbf{H}) \times \mathbf{H}$$

$$\nabla \cdot \mathbf{V} = 0, \qquad \nabla \cdot \mathbf{H} = 0 \tag{5.4.1-1}$$

$$\nabla \times \mathbf{H} = \sigma (\mathbf{E} + \mathbf{V} \times \mu\mathbf{H}), \qquad \nabla \times \mathbf{E} = 0$$

This flow is clearly axisymmetric, the axis of symmetry being the $0x$ axis passing through the center of the sphere 0 and being parallel to $\mathbf{V}_\infty$. The velocity and magnetic field are in the meridian plane. The position of a point M will be given by the spherical coordinates r, θ, φ, where $r = 0M$, $\theta = (0x, 0M)$, and φ defines the position of the half plane $(0x, M)$. The components of the magnetic field on the axis associated with the spherical coordinates, having $\mathbf{i}_1, \mathbf{i}_2, \mathbf{i}_3$ for unit vectors, are H_r, H_θ, $H_\varphi = 0$; we have

$$\nabla \times \mathbf{H} = -\frac{\mathbf{i}_3}{r} \left\{ \frac{\partial H_r}{\partial \theta} - \frac{\partial (rH_\theta)}{\partial r} \right\}$$

From the next to the last of Eqs. (5.4.1.1), we have $\mathbf{E} = \mathbf{i}_3 E_\varphi$, and from the last equation, $E_\varphi \sin\theta = \text{const}$; since E_φ is zero on the $0x$ axis ($\theta = 0$), the electric field is zero throughout.

We introduce a reference magnetic field $h = M/a^3$ and we write Eqs. (5.4.1.1) in dimensionless form. To do this, we divide lengths by a, and divide $\mathbf{H}$, $\mathbf{V}$, and p by h, V_∞, and $\rho_0 V_\infty^2$, respectively; we write

$$P = \frac{p}{\rho_0 V_\infty^2} + \frac{1}{2} \frac{V^2}{V_\infty^2}$$

keeping the same notation for the dimensionless magnitudes, the dimensionless relations become

$$(\nabla \times \mathbf{V}) \times \mathbf{V} + \nabla P = \beta (\nabla \times \mathbf{H}) \times \mathbf{H}$$

$$\nabla \cdot \mathbf{V} = 0, \qquad \nabla \cdot \mathbf{H} = 0 \tag{5.4.1-2}$$

$$\nabla \times \mathbf{H} = R_{\mathrm{m}} \mathbf{V} \times \mathbf{H}$$

with

$$\beta = \mu h^2 / \rho_0 V_\infty^2, \qquad R_{\mathrm{m}} = V_\infty a \sigma \mu$$

R_{m} is called the magnetic Reynolds number.

The dimensional magnetic field, proportional to h, tends to zero with β as $\sqrt{\beta}$. In the absence of magnetic field ($\beta = 0$), the solution of the equation (5.4.1-2) which represents uniform flow past a sphere of unit radius is well known:

$$\mathbf{V} = \mathbf{V}_0(r, \theta) = \mathbf{i}_1(1 - 1/r^3)\cos\theta - \mathbf{i}_2(1 + 1/2r^3)\sin\theta$$

$$P = \text{const}, \qquad \mathbf{H} = \mathbf{H}_0(r, \theta)$$

$$p = p_0(r, \theta) = P - \tfrac{1}{2}V_0^2(r, \theta)$$

For small values of the magnetic field (β small), we look for a solution of (5.4.1-2) in the following form:

$$\begin{aligned} \mathbf{V} &= \mathbf{V}_0(r, \theta) + \beta\mathbf{V}_1(r, \theta) + \cdots \\ p &= p_0(r, \theta) + \beta p_1(r, \theta) + \cdots \\ \mathbf{H} &= \mathbf{H}_0(r, \theta) + \cdots \end{aligned} \tag{5.4.1-3}$$

Substituting these values in the equations of motion and equating coefficients of corresponding powers of β, we get the following equations:

$$\nabla \cdot \mathbf{H}_0 = 0, \qquad \nabla \times \mathbf{H}_0 = R_m \mathbf{V}_0 \times \mathbf{H}_0 \tag{5.4.1-4}$$

$$\begin{aligned} (\nabla \times \mathbf{V}_1) \times \mathbf{V}_0 + \nabla(p_1 + \mathbf{V}_0 \cdot \mathbf{V}_1) &= (\nabla \times \mathbf{H}_0) \times \mathbf{H}_0 \\ \nabla \cdot \mathbf{V}_1 &= 0 \end{aligned} \tag{5.4.1-5}$$

5.4.2. Calculation of the Magnetic Field

For small values of the parameter β, the dimensionless magnetic field reduces to $\mathbf{H}_0(r, \theta)$ and the dimensional magnetic field becomes

$$(\beta\rho_0 V_0^2/\mu)^{1/2}\,\mathbf{H}_0(r, \theta)$$

The field $\mathbf{H}_0(r, \theta)$ is determined by Eqs. (5.4.1-4) and by the following boundary conditions:

(a) Near the center of the sphere, the magnetic field is that due to a dipole of moment M and axis $0x$:

$$H_{0r} = (2\cos\theta)/r^3, \qquad H_{0\theta} = (2\sin\theta)/r^3, \qquad H_0 = 0$$

(b) On the surface of the sphere, the normal component of the induction is continuous and so is the tangential component of the magnetic field.

(c) At infinity, the magnetic field goes to zero.

It is not possible to satisfy all these conditions exactly, but it is possible to satisfy them to a good approximation. From the first of Eqs. (5.4.1-5) there exists a function $A(r, \theta)$ such that

$$H_{0r} = \frac{1}{r^2 \sin\theta} \frac{\partial A}{\partial \theta}, \qquad H_{0\theta} = -\frac{1}{r \sin\theta} \frac{\partial A}{\partial r} \tag{5.4.2-1}$$

The second equation of (5.4.1-4) becomes

$$\mathscr{L}(A) = R_{\mathrm{m}} \left\{ \left(1 - \frac{1}{r^3}\right) \cos\theta \frac{\partial A}{\partial r} - \frac{1}{r}\left(1 + \frac{1}{2r^3}\right) \sin\theta \frac{\partial A}{\partial \theta} \right\} \tag{5.4.2-2}$$

where

$$\mathscr{L}(A) \equiv \frac{\partial^2 A}{\partial r^2} + \frac{\sin\theta}{r^2} \frac{\partial}{\partial \theta} \left(\frac{1}{\sin\theta} \frac{\partial A}{\partial \theta} \right)$$

We solve Eq. (5.4.2-2) assuming the magnetic Reynolds number to be small and expanding the function $A(r, \theta)$ in increasing powers of this number:

$$A(r, \theta) = A_0(r, \theta) + R_{\mathrm{m}} A_1(r, \theta) + \cdots$$

Comparing both sides of Eq. (5.4.2-2), we get

$$\mathscr{L}(A_0) = 0$$

$$\mathscr{L}(A_1) = \left(1 - \frac{1}{r^3}\right) \cos\theta \frac{\partial A_0}{\partial r} - \frac{1}{r}\left(1 + \frac{1}{2r^3}\right) \sin\theta \frac{\partial A_0}{\partial \theta}$$

The function $A_0(r, \theta)$ can be expressed in the form

$$A_0(r, \theta) = \sum_{n=1}^{\infty} \frac{a_n}{r^n} \sin^2\theta \, P_n'(\cos\theta)$$

where

$$P_n(z) = \frac{1}{2^n \, n!} \frac{d^n}{dz^n} (z^2 - 1)^n$$

is the Legendre polynomial of order n, $P_n'(z)$ its derivative, and the coefficients a_n are arbitrary constants. The function $A_1(r, \theta)$ can then be expressed in the form

$$A_1(r, \theta) = \sum_{n=1}^{\infty} \frac{b_n}{r^n} \sin^2\theta \, P_n'(\cos\theta) + A_1^*(r, \theta) + A_1^{**}(r, \theta)$$

The series represents the general solution of the homogeneous equation $\mathscr{L}(A_1) = 0$, while the functions A_1^* and A_1^{**} are the particular solutions of the equations

$$\mathscr{L}(A_1^*) = \cos\theta \frac{\partial A_0}{\partial r} - \frac{\sin\theta}{r} \frac{\partial A_0}{\partial \theta}$$

$$\mathscr{L}(A_1^{**}) = -\frac{1}{r^3} \left\{ \cos\theta \frac{\partial A_0}{\partial r} + \frac{\sin\theta}{2r} \frac{\partial A_0}{\partial \theta} \right\}$$

We can take $A_1^* = \frac{1}{2}(r \cos\theta)\, A_0(r, \theta)$; and, to solve the second equation, we assume that, in the series for $A_0(r, \theta)$, all the coefficients a_1, a_2,... are zero except the coefficient a_n. We then have

$$\mathscr{L}(A_1^{**}) = \frac{a_n}{r^{n+4}} \left\{ n \cos\theta \sin^2\theta \, P_n' - \frac{\sin\theta}{2} \frac{d}{d\theta} (\sin^2\theta \, P_n') \right\}$$

or

$$A_1^{**} = a_n \alpha_n(\theta)/r^{n+2}$$

where $\alpha_n(\theta)$ is a function of θ alone, which can be determined. In the general case, we have

$$A_1^{**} = \sum_{n=1}^{\infty} a_n \alpha_n(\theta)/r^{n+2}$$

If we then assume that all the coefficients a_n and b_n are zero except for a_1 and b_2, we obtain the following expression for $A(r, \theta)$ to represent the magnetic field outside the sphere:

$$A_e = a_1 \frac{\sin^2\theta}{r} + R_m \left(\frac{a_1}{2} + \frac{3b_2}{r^2} \right) \sin^2\theta \cos\theta \qquad (5.4.2\text{-}3)$$

Inside the sphere, the electric field is still zero and the velocity is zero also ($\mathbf{V} = 0$); we then have $\nabla \times \mathbf{H} = 0$, namely

$$\mathscr{L}(A) \equiv \frac{\partial^2 A}{\partial r^2} + \frac{\sin\theta}{r^2} \frac{\partial}{\partial \theta} \left(\frac{1}{\sin\theta} \frac{\partial A}{\partial \theta} \right] = 0$$

The solution of this equation, which reduces to $(\sin^2\theta)/r$ in the neighborhood of the center of the sphere, is

$$A = \frac{\sin^2\theta}{r} + c_1 r \cos\theta + \sum_{n=2}^{\infty} c_n r^n \sin^2\theta \, P'_{n-1}(\cos\theta)$$

To complete the calculation of the magnetic field $\mathbf{H}_0(r, \theta)$, we still have to satisfy the conditions on the surface of the sphere—continuity of normal magnetic induction μH_r and tangential magnetic field H_θ; in other words,

$$\mu \frac{\partial A_e}{\partial \theta}(1, \theta) \equiv \mu' \frac{\partial A_i}{\partial \theta}(1, \theta), \qquad \frac{\partial A_e}{\partial r}(1, \theta) \equiv \frac{\partial A_i}{\partial r}(1, \theta) \tag{5.4.2-4}$$

These boundary conditions, which are identities in θ, can be satisfied if we take as the function $A(r, \theta)$ the expression defined by Eq. (5.4.2-3) in the region outside the sphere and the value A_i defined as follows in the region inside the sphere:

$$A_i = \frac{\sin^2\theta}{r} + c_2 r^2 \sin^2\theta + c_3 r_3 \sin^2\theta \cos\theta \tag{5.4.2-5}$$

The boundary conditions (5.4.2-4) then reduce to the equations

$$\mu a_1 - \mu' c_2 = 0, \qquad a_1 + 2c_2 = 0$$

$$\mu R_m(a_1 + 2b_2) - 2\mu' c_3 = 0, \qquad 2b_2 R_m + c_3 = 0$$

which give

$$a_1 = \frac{3\mu'}{2\mu + \mu'}, \qquad b_2 = \frac{9\mu\mu'}{2(2\mu + \mu')(3\mu + 2\mu')}$$

$$c_2 = \frac{\mu - \mu'}{2\mu + \mu'}, \qquad c_3 = \frac{3\mu\mu' R_m}{(2\mu + \mu')(3\mu + 2\mu')}$$

In practice, the magnetic permeability of the fluid μ is equal to 1, the magnetic permeability of the sphere μ' is either equal to 1 (paramagnetic sphere) or to infinity (ferromagnetic sphere).

These results determine the magnetic field to the order $O(R_m)$; but calculations to $O(R_m^2)$ have been carried out by M. Bois. The lines of force of the magnetic field to $O(R_m^2)$ are shown in Fig. 5.4.2.a, for $\mu = \mu'$ and $R_m = 1/2$.

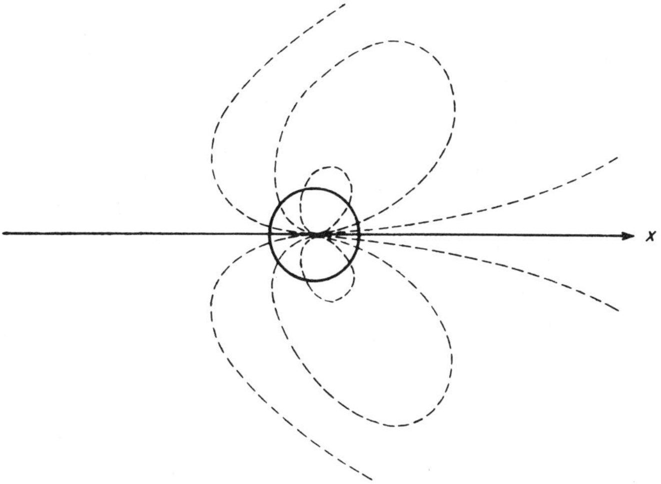

FIG. 5.4.2.a. Lines of force of the magnetic field; $\mu = \mu'$, $R_m = 1/2$.

5.4.3. COMPUTATION OF THE VELOCITY

To compute the velocity to the same approximation, it is sufficient to determine $\mathbf{V}_1(r, \theta)$. Since the divergence of this vector is zero, there exists a function $\Psi_1(r, \theta)$ such that

$$V_{1r} = \frac{1}{r^2 \sin\theta} \frac{\partial \Psi}{\partial \theta}, \qquad V_{1\theta} = -\frac{1}{r \sin\theta} \frac{\partial \Psi}{\partial r} \tag{5.4.3-1}$$

We have $\nabla \times \mathbf{V}_1 = \mathbf{i}_3\, \omega$, with

$$\omega = \frac{1}{r}\left\{\frac{\partial(rV_{1\theta})}{\partial r} - \frac{\partial V_{1r}}{\partial \theta}\right\} = -\frac{\mathscr{L}(\Psi)}{r \sin\theta}$$

We eliminate the function p_1, taking the curl of both sides in the first of Eqs. (5.4.1-5):

$$\nabla \times \{(\nabla \times \mathbf{V}_1) \times \mathbf{V}_0\} = \nabla \times \{(\nabla \times \mathbf{H}_0) \times \mathbf{H}_0\}$$

or, writing

$$\tilde{\omega} = -\frac{1}{r^2 \sin^2\theta} \mathscr{L}(\Psi)$$

we have

$$\left(1 - \frac{1}{r^3}\right) \cos\theta \frac{\partial \tilde{\omega}}{\partial r} - \frac{1}{r}\left(1 + \frac{1}{2r^3}\right) \sin\theta \frac{\partial \tilde{\omega}}{\partial \theta} = f(r, \theta; R_{\mathrm{m}}) \quad (5.4.3\text{-}2)$$

with

$$f(r, \theta; R_{\mathrm{m}}) = \frac{1}{r^2 \sin\theta}\left\{\frac{\partial A}{\partial r}\frac{\partial}{\partial \theta}\left[\frac{\mathscr{L}(A)}{r^2 \sin^2\theta}\right] - \frac{\partial A}{\partial \theta}\frac{\partial}{\partial r}\left[\frac{\mathscr{L}(A)}{r^2 \sin^2\theta}\right]\right\}$$

We solve Eq. (5.4.3-2) by expanding the unknown function $\omega = r \sin\theta\, \tilde{\omega}$, and the known right-hand side, as series of derivatives of Legendre polynomials:

$$\omega(r, \theta) = \sum_{n=1}^{\infty} \sin\theta\, \omega_n(r)\, P_n'(\cos\theta)$$

$$f(r, \theta; R_{\mathrm{m}}) = \sum_{n=1}^{\infty} f_n(r)\, P_n'(\cos\theta)$$

The Legendre polynomials satisfy the following differential equation and recurrence formulas:

$$\frac{d}{d\theta}\{\sin^2\theta\, P_n'(\cos\theta)\} - n(n+1) \sin\theta\, P_n(\cos\theta) = 0$$

$$(2n+1)\, P_n(\cos\theta) = P_{n+1}'(\cos\theta) - P_{n-1}'(\cos\theta)$$

$$\cos\theta\, P_n'(\cos\theta) = P_{n-1}'(\cos\theta) + nP_n(\cos\theta)$$

The derivatives of the polynomials P_n' form an infinite sequence of linearly independent polynomials; it is therefore possible to substitute the above series for $f(r, \theta; R_{\mathrm{m}})$ and $\omega(r, \theta)$ and compare coefficients of $P_n'(\cos\theta)$ on both sides of the equation term by term. We then get an infinite number of differential equations in $\omega_n(r)$:

$$\frac{n+2}{2n+3}\left\{\left(r - \frac{1}{r^2}\right)\omega_{n+1}' + \left(n + 2 + \frac{n+5}{2r^3}\right)\omega_{n+1}\right\}$$
$$+ \frac{n-1}{2n-1}\left\{\left(r - \frac{1}{r^2}\right)\omega_{n-1}' - \left(n - 1 + \frac{n-4}{2r^3}\right)\omega_{n-1}\right\} = rf_n(r) \quad (5.4.3\text{-}3)$$

In particular, for $n = 1$, we get

$$\left(r - \frac{1}{r^2}\right)\omega_2' + 3\left(1 + \frac{1}{r^3}\right)\omega_2 = \frac{5}{3} r f_1(r)$$

or

$$\omega_2(r) = \frac{5}{3} \frac{r^3}{(r^3 - 1)^2} \int_{r_0}^{r} (\alpha^3 - 1) f_1(\alpha)\, d\alpha$$

The lower limit of the integral is arbitrary, but since the vorticity is finite at the surface of the sphere $r = 1$, we have to take $r_0 = 1$.

Once the function $\omega(r, \theta)$ is known, it is easy to calculate $\Psi(r, \theta)$ and then the velocity. We have

$$\Psi(r, \theta) = \sum_{n=1}^{\infty} \Psi_n(r) \sin^2\theta\, P_n'(\cos\theta)$$

with

$$\Psi_n'' - \frac{n(n+1)}{r^2} \Psi_n = -r\, \omega_n(r)$$

or

$$\Psi_n = Ar^{n+1} + Br^{-n} + \frac{1}{2n+1} \int_1^r \left\{ \frac{\xi^{n+2}}{r^n} - \frac{r^{n+1}}{\xi^{n-1}} \right\} \omega_n(\xi)\, d\xi$$

The integration constants A and B have a zero sum because the sphere is a streamline: $\Psi_n(1) = 0$; also, $\Psi_n(r)$ has to be bounded at infinity; hence,

$$A = \frac{1}{2n+1} \int_1^{\infty} \frac{\omega_n(\xi)}{\xi^{n-1}}\, d\xi$$

In particular, we have

$$5\, \Psi_2 = \left(r^3 - \frac{1}{r^2}\right) \int_1^{\infty} \frac{\omega_2(\xi)}{\xi}\, d\xi + \int_1^{r} \left(\frac{\xi^4}{r^2} - \frac{r^3}{\xi}\right) \omega_2(\xi)\, d\xi$$

$$\Psi_2'(1) = \int_1^{\infty} \frac{\omega_2(\xi)}{\xi}\, d\xi = \tfrac{5}{9} \int_1^{\infty} f_1(\alpha)\, d\alpha$$

The components of the vector $\mathbf{V}_1(r, \theta)$ are then given by the formulae

$$V_{1r} = \frac{1}{r^2} \sum_{n=1}^{\infty} n(n+1)\, \Psi_n(r)\, P_n(\cos\theta)$$

$$V_{1\theta} = -\frac{1}{r} \sum_{n=1}^{\infty} \Psi_n'(r) \sin\theta\, P_n'(\cos\theta)$$

5.4.4. Calculation of the Drag

The force exerted by the fluid on the sphere is, by symmetry, along the axis $0x$. This force results from (1) a dynamic part

$$\iint_{(S)} \mathbf{n}\, p\, dS = -2\pi a^2\, \mathbf{x} \int_0^{\pi} p \sin\theta \cos\theta\, d\theta$$

where $p = p(a, \theta)$ is the pressure on the sphere, and (2) a magnetic force, the reaction to the Lorentz forces on each fluid particle:

$$\iiint \mathbf{J} \times \mu\mathbf{H}\, d\tau = \iint_{(S)} \mathsf{H} \cdot \mathbf{n}\, dS$$

where

$$\mathsf{H} = \mu(\mathbf{H}, \mathbf{H}) - \tfrac{1}{2}\mu H^2\, \mathsf{U}.$$

To compute the force of dynamic origin, we write

$$p = \rho_0 V_\infty^2 \{p_0(r, \theta) + \beta\, p_1(r, \theta) + \cdots\}$$

the term $p_0(r, \theta)$ does not contribute to the drag because, in the absence of a magnetic field, the drag is zero by d'Alembert's paradox. Writing $p_1 + \mathbf{V}_0 \cdot \mathbf{V}_1 = P_1$, we have, by projecting the first of Eqs. (5.4.1-5) along the vector $\mathbf{i}_1$,

$$\frac{\partial P_1}{\partial\theta} + \frac{\mathscr{L}(A)}{r^2 \sin^2\theta} \frac{\partial A}{\partial\theta} + \left(r - \frac{1}{r^2}\right) \omega \cos\theta = 0$$

or

$$P_1(1, \theta) + \int^{\theta} \frac{\mathscr{L}(A)}{\sin^2\theta} \frac{\partial A}{\partial\theta}\, d\theta = \text{const}$$

We write

$$P_1(1, \theta) = -\sum_{n=1}^{\infty} g_n(1)\, P_n(\cos\theta)$$

where $P_n(z)$ is the Legendre polynomial of order n. The coefficient of aerodynamic drag C_p is equal to the ratio of the drag to $\rho_0 V_\infty^2 a^2$:

$$C_p = -2\pi\beta \int_0^{\pi} p_1(1, \theta) \sin\theta \cos\theta\, d\theta$$

Expanding $p_1(1, \theta)$ in a series of Legendre polynomials and using the orthogonality relations satisfied by these polynomials

$$\int_{-1}^{1} P_m(z)\, P_n(z)\, dz = \frac{2}{2n+1}\,\delta_{m,n}$$

$$\int_{-1}^{1} (1 - z^2)\, P_m'(z)\, P_n'(z)\, dz = \frac{2}{2n+1}\,\frac{(n+1)!}{(n-1)!}\,\delta_{m,n}$$

where $\delta_{m,n} = 0$ for $m \neq n$, and $\delta_{n,n} = 1$, we get

$$C_p = 2\pi\beta\,\{\tfrac{2}{3}\, g_1(1) + \tfrac{6}{5}\,\Psi_2'(1)\}$$

With the approximation used in the previous section to represent $A(r, \theta)$ inside the sphere, we have

$$f(r, \theta; R_\mathrm{m}) = -3R_\mathrm{m} a_1^2\,\frac{1 + 7\cos^2\theta}{r^7} + O(R_\mathrm{m}^2)$$

$$P_1(1, \theta) = \mathrm{const} - 2R_\mathrm{m} a_1^2 \cos^3\theta + O(R_\mathrm{m}^2)$$

Hence, to order $O(R_\mathrm{m})$,

$$f_1(r) = -\tfrac{36}{5} R_\mathrm{m}(a_1^2/r^7), \qquad \Psi_2'(1) = -\tfrac{2}{3} R_\mathrm{m} a_1^2$$

$$g_1(1) = \tfrac{6}{5} R_\mathrm{m} a_1^2$$

It follows that $C_p = \beta O(R_\mathrm{m}^2)$.

The drag coefficient of magnetic origin C_m has the value

$$C_\mathrm{m} = \frac{1}{\rho_0 V_\infty^2 a^2} \iint_{(S)} (\mathbf{H} \cdot \mathbf{n}) \cdot \mathbf{x}\, dS$$

$$(\mathbf{H} \cdot \mathbf{r}) \cdot \mathbf{x} = \mu H_r(\cos\theta\, H_r - \sin\theta\, H_\theta) - \frac{\mu H^2}{2}\cos\theta$$

$$= \beta\rho_0 V_\infty^2\,\frac{\cos\theta(\partial A/\partial\theta)^2 + 2r\sin\theta(\partial A/\partial\theta)(\partial A/\partial r) - r^2\cos\theta(\partial A/\partial r)^2}{2r^4\sin^2\theta}$$

Replacing dS by its value $2\pi a^2 \sin\theta\, d\theta$, we obtain for the magnetic drag coefficient

$$C_\mathrm{m} = \tfrac{8}{5}\pi\beta a_1^2 R_\mathrm{m} + O(R_\mathrm{m}^2)$$

Hence, the magnetic drag is more important than the aerodynamic drag. Substituting the value of a_1, we get the relations for the total drag coefficient:

$$C_x = \left\{\left(\iint \mathbf{n}p\, dS + \iiint \mathbf{J}\mathbf{x}\mu\mathbf{H}\, d\tau\right)\Big/\frac{1}{2}\rho V_\infty{}^2\pi a^2\right\}\cdot\mathbf{x}$$

$$C_x = \{12\mu'/(2\mu+\mu')\}^2\,\beta R_{\mathrm{m}}/5$$

The computation of the following approximation was performed by M. Bois (1970); the fluid is assumed paramagnetic ($\mu = 1$):

$$C_x = \frac{16}{5}\beta R_{\mathrm{m}}\left(1 - \frac{101{,}089}{323{,}400}R_{\mathrm{m}}{}^2\right) \qquad \text{for} \quad \mu' = 1$$

$$C_x = \frac{144}{5}\beta R_{\mathrm{m}}\left(1 - \frac{1017}{1960}R_{\mathrm{m}}{}^2\right) \qquad \text{for} \quad \mu' = \infty$$

5.5. Flow Past a Sphere with a Cavity

5.5.1. Equations of the Problem

In this section, we consider the steady flow of a viscous incompressible fluid past a sphere with a cavity in the presence of a magnetic field. At infinity, the velocity of the fluid V_∞ and the magnetic field H_∞ are uniform and parallel (Fig. 5.5.1.a). The

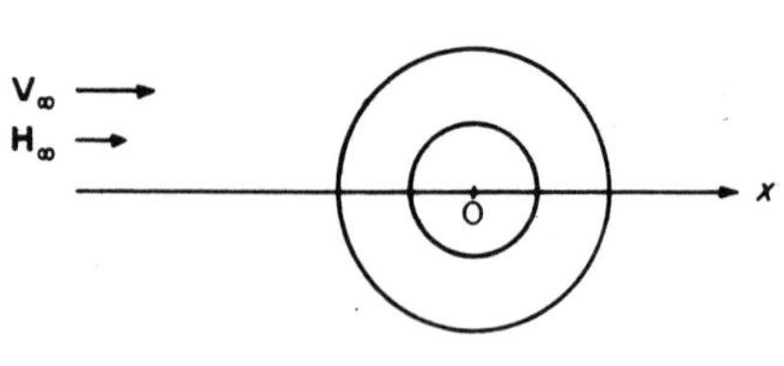

FIG. 5.5.1.a. Fluid velocity and magnetic field at infinity.

equations of motion

$$\rho_0\mathbf{V}\cdot\nabla\mathbf{V} + \nabla p = \rho_0\,\nu\nabla^2\mathbf{V} + (\mathbf{V}\times\mathbf{H})\times\mu\mathbf{H}$$

$$\nabla\cdot\mathbf{V} = 0, \qquad \nabla\cdot\mathbf{H} = 0 \tag{5.5.1-1}$$

$$\nabla\times\mathbf{H} = \sigma(\mathbf{E} + \mathbf{V}\times\mathbf{B}), \qquad \nabla\times\mathbf{E} = 0$$

can be written in dimensionless form. To do this, we divide **H, V,** and p, respectively, by H_∞, V_∞, and $a/\rho_0\nu V_\infty$, and x by a (the larger radius). Keeping the same notation for simplicity, we get the following dimensionless equations:

$$R\,\mathbf{V}\cdot\nabla\mathbf{V} + \nabla p = \nabla^2\mathbf{V} + M^2(\mathbf{V}\times\mathbf{H})\times\mathbf{H}$$

$$\nabla\cdot\mathbf{V} = 0, \qquad \nabla\cdot\mathbf{H} = 0 \tag{5.5.1-2}$$

$$\nabla\times\mathbf{H} = R_{\mathrm{m}}(\mathbf{V}\times\mathbf{H})$$

where $R = V_\infty a/\nu$ is the Reynolds number, $R_{\mathrm{m}} = V_\infty a\sigma$ is the magnetic Reynolds number, and $M = aH_\infty(\sigma/\rho_0)^{1/2}$ is the Hartman number. Since the motion is axisymmetric, the electric field is zero if it is to have a finite value on $0x$.

Assuming the magnetic Reynolds number to be negligible, the magnetic field is independent of the fluid velocity, and will have the same value as in the static case, that is, outside the sphere ($r \geqslant 1$),

$$\mathbf{H} = \mathbf{H}_0 = \nabla\left\{\left(r - \frac{\gamma}{r^2}\right)\cos\theta\right\} \tag{5.5.1-3}$$

with

$$\gamma = \frac{(2\mu' + \mu)(\mu' - \mu)[1 - (b^3/a^3)]}{(2\mu' + \mu)(\mu' + 2\mu) - 2(b^3/a^2)(\mu' - \mu)^2}$$

In the spherical shell ($b/a \leqslant r \leqslant 1$),

$$\mathbf{H} = \mathbf{H}_1 = \nabla\left\{\left(\beta r + \frac{\beta'}{r^2}\right)\cos\theta\right\}$$

In the cavity ($0 \leqslant r \leqslant b/a$),

$$\mathbf{H} = \mathbf{H}_2 = -\nabla(\delta r\cos\theta)$$

with

$$\delta = \frac{-9\mu\mu'}{(2\mu' + \mu)(\mu' + 2\mu) - 2(b^3/a^3)(\mu' - \mu)^2}$$

$$\beta\frac{b^3}{a^3} + \beta' + \delta\frac{b^3}{a^3} = 0$$

$$\mu\beta\frac{b^3}{a^3} - 2\mu\beta' + \mu'\delta\frac{b^3}{a^3} = 0$$

where μ is the magnetic permeability of the fluid and the cavity, μ' that of the spherical shell, a and b are the external and internal radii of the spherical shell, and r and θ are dimensionless polar coordinates with origin at the center of the spherical shell.

Evidently, the magnetic field outside the sphere is the sum of a uniform magnetic field H_∞ and the field of a dipole of moment $H_\infty \gamma a^3$ directed along $0x$. In the cavity, there exists a uniform field of intensity inversely proportional to μ'.

If we now assume that the Reynolds number is also small, we have to solve the following equations to compute the velocity and pressure:

$$\nabla p = \nabla^2 \mathbf{V} + M^2(\mathbf{V} \times \mathbf{H}_0) \times \mathbf{H}_0), \qquad \nabla \cdot \mathbf{V} = 0 \qquad (5.5.1\text{-}4)$$

with the boundary conditions

$$\mathbf{V}(\infty, \theta) = \mathbf{x}, \qquad \text{condition at infinity}$$
$$\mathbf{V}(1, \theta) = 0, \qquad \text{condition on the sphere}$$

5.5.2. Computation of the Velocity

We can solve the problem of finding the velocity field by the method of double asymptotic expansions. We write, in the neighborhood of the sphere,

$$\mathbf{V}(r, \theta) = \mathbf{V}_0(r, \theta) + M\mathbf{V}_1(r, \theta) + M^2\mathbf{V}_2(r, \theta) + \cdots$$
$$p(r, \theta) = p_0(r, \theta) + Mp_1(r, \theta) + M^2p_2(r, \theta) + \cdots$$

and far from the sphere,

$$\mathbf{V}(r, \theta) = \mathbf{W}_0(\bar{r}, \theta) + M\mathbf{W}_1(\bar{r}, \theta) + M^2\mathbf{W}_2(\bar{r}, \theta) + \cdots$$
$$p(r, \theta) = \pi_0(\bar{r}, \theta) + M\pi_1(\bar{r}, \theta) + M^2\pi_2(\bar{r}, \theta) + \cdots$$

where $\bar{r} = Mr$, $\bar{x} = Mx$.

Substituting the above expansions in (5.5.1-4), we get the following system:

$$\nabla \cdot \mathbf{V}_i = 0, \qquad \nabla p_i - \nabla^2 \mathbf{V}_i = 0, \qquad \text{for} \quad i = 0, 1$$

$$\nabla \cdot \mathbf{V}_i = 0, \qquad \nabla p_i - \nabla^2 \mathbf{V}_i = (\mathbf{V}_{i-2} \times \mathbf{U}_0) \times \mathbf{U}_0\,, \qquad \text{for} \quad i > 2$$

$$\nabla \cdot \mathbf{W}_i = 0, \qquad \nabla \pi_{i+1} - \nabla^2 \mathbf{W}_i - (\mathbf{W}_i \times \mathbf{x}) \times \mathbf{x} = 0, \qquad \text{for} \quad i = 0, 1, 2$$

$$\nabla \cdot \mathbf{W}_i = 0, \qquad \nabla \pi_{i+1} - \nabla^2 \mathbf{W}_i - (\mathbf{W}_i \times \mathbf{x}) \times \mathbf{x}$$
$$= (\mathbf{W}_{i-2} \times \mathbf{U}_0') \times \mathbf{x} + (\mathbf{W}_{i-2} \times \mathbf{x}) \times \mathbf{U}_0' \qquad \text{for} \quad i \geqslant 3$$

where we have written $\mathbf{H}_0(r, \theta) = \mathbf{x} + M^3\mathbf{H}_0'(\bar{r}, \theta)$

We apply the boundary conditions at infinity only to the first expansion and those at infinity to the second expansion,

$$\mathbf{V}_i(1, \theta) = 0, \qquad \text{for} \quad i = 0, 1, 2, \ldots$$

$$\mathbf{W}_0(\infty, \theta) = \mathbf{x}$$

$$\mathbf{W}_i(\infty, \theta) = 0, \qquad \text{for} \quad i = 1, 2, \ldots$$

At each step in the integration, certain constants appear, which result from the fact that only a part of the boundary conditions is imposed on each expansion. We determine the constants by matching the inner and outer expansions, i.e., we require that the expansion near the sphere and that near infinity be identical term-by-term for $r = O(M^{-\alpha})$, $0 < \alpha < 1$, the order of the intermediate region for which both expressions are valid.

a. Computation of $\mathbf{W}_0(\bar{r}, \theta)$

In outer variables $(\bar{r}, \theta)$, the obstacle $\bar{r} = M$ reduces to a point for $M = 0$, and the flow is not disturbed. We then have $\mathbf{W}_0(\mathbf{r}, \theta) = \mathbf{x}$.

b. Computation of $\mathbf{V}_0(r, \theta)$

The vector $\mathbf{V}_0(r, \theta)$ must satisfy $\nabla \cdot \mathbf{V}_0 = 0$ and $\nabla p_0 - \nabla^2 \mathbf{V}_0 = 0$, the condition $\mathbf{V}_0(1, \theta) = 0$, and match $\mathbf{W}_0(r, \theta) = \mathbf{x}$. The solution of Stokes' problem satisfies the first two conditions; the corresponding vector $\mathbf{V}_0(r, \theta)$ is derived from the stream function:

$$\Psi_0(r, \theta) = \frac{r^2 \sin^2\theta}{2} \left(1 - \frac{3}{2r} + \frac{1}{2r^3}\right)$$

In intermediate variables (r_α, θ), where $r_\alpha = M^\alpha r$, $0 < \alpha < 1$, we have

$$V_{0r}(r_\alpha, \theta, M) = \cos\theta \left(1 - \frac{3M^\alpha}{2r_\alpha}\right) + O(M^{3\alpha})$$

$$V_{0\theta}(r_\alpha, \theta, M) = -\sin\theta \left(1 - \frac{3M^\alpha}{4r_\alpha}\right) + O(M^{3\alpha})$$

which matches $\mathbf{W}_0(r, \theta) = \mathbf{x}$ to order $O(1)$ for any $\alpha > 0$.

c. Computation of $\mathbf{W}_1(\bar{r}, \theta)$

We seek the solution of the system satisfied by $\mathbf{W}_1$ and π_2 in the form

$$\mathbf{W}_1(\bar{r}, \theta) = [\exp(\bar{x})]\, \bar{\nabla}\Phi_1 + [\exp(-\bar{x})]\, \bar{\nabla}\Phi_2$$

$$\pi_2(\bar{r}, \theta) = [\exp(\bar{x})]\, \partial\Phi_1/\partial\bar{x} - [\exp(-\bar{x})]\, \partial\Phi_2/\partial\bar{x}$$

The equations satisfied by $\mathbf{W}_1$ and π_2 reduce to

$$\bar{\nabla}^2\Phi_1 + \partial\Phi_1/\partial\bar{x} = 0, \qquad \bar{\nabla}^2\Phi_2 - \partial\Phi_2/\partial\bar{x} = 0$$

The functions Φ_1 and Φ_2 can be written in the form of series of successive derivatives of the fundamental solutions:

$$\Phi_1 = (1/\bar{r}) \exp[-\tfrac{1}{2}(\bar{r} + \bar{x})], \qquad \Phi_2 = (1/\bar{r}) \exp[-\tfrac{1}{2}(\bar{r} - \bar{x})]$$

The solution

$$\begin{aligned}\mathbf{W}_1(\bar{r}, \theta) = C_0(&[\exp(\bar{x})]\, \bar{\nabla}\{(1/\bar{r}) \exp[-\tfrac{1}{2}(\bar{r} + \bar{x})]\} \\ &-[\exp(-\bar{x})]\, \bar{\nabla}\{(1/\bar{r}) \exp[-\tfrac{1}{2}(\bar{r} - \bar{x})]\})\end{aligned}$$

is symmetric with respect to the x axis, and satisfies the equations of the problem and the conditions at infinity. In intermediate variables, $\mathbf{W}_0(\bar{r}, \theta) + M\mathbf{W}_1(\bar{r}, \theta)$ gives

$$V_r = \cos\theta - 2C_0M^\alpha \frac{\cos\theta}{r_\alpha} + C_0M \frac{\cos\theta}{2} + O(M^{2-\alpha})$$

$$V_\theta = -\sin\theta + C_0M^\alpha \frac{\sin\theta}{r_\alpha} - C_0M \frac{\sin\theta}{2} + O(M^{2-\alpha})$$

which, if we take $C_0 = 3/4$, matches $\mathbf{V}_0(r, \theta)$ to order $O(M)$ and for any value of α between 0 and 1.

d. Computation of $\mathbf{V}_1(r, \theta)$

The first term of the outer expansion which does not match the inner solution is of order $O(M)$. The function $\mathbf{V}_1(r, \theta)$ must satisfy the same equations as those for $\mathbf{V}_0(r, \theta)$, the condition $\mathbf{V}_1(1, \theta) = 0$, and the conditions for matching.

The function $V_1(r, \theta) = kV_0(r, \theta)$ satisfies the first two conditions. The components of $V_0 + MV_1$ expressed in intermediate variables are then

$$V_r = \cos\theta \left\{ 1 - \frac{3M^\alpha}{2r^\alpha} + kM - kM^{1+\alpha}\frac{3}{2r^\alpha} \right\} + O(M^{3\alpha})$$

$$V_\theta = -\sin\theta \left\{ 1 - \frac{3M^\alpha}{4r^\alpha} + kM - kM^{1+\alpha}\frac{3}{4r^\alpha} \right\} + O(M^{3\alpha})$$

which, if we take $k = C_0/2 = 3/8$, matches $\mathbf{W}_0 + M\mathbf{W}_1$ to order $O(M)$ for any value of α between 1/3 and 1.

The method can be continued further; only the complexity of computation will require us to stop at some point.

5.5.3. Computation of the Drag

When the velocity field is known, the force imposed on the sphere by the fluid is calculated as in Section 5.4. The force F will be along $0x$ by symmetry and will have the magnitude

$$F = \rho_0 \nu V_\infty a(C_p + C_{\text{m}})$$

The coefficient C_p corresponding to the dynamic force has the value

$$C_p = \frac{1}{\rho_0 \nu V_\infty a} \iint_{(S)} (\mathbf{T} \cdot \mathbf{n}) \cdot \mathbf{x}\, dS$$
$$= 2\pi \int_0^\pi \left\{ -p\cos\theta + V_r + \frac{\partial V_r}{\partial r} + \frac{\partial(\mathbf{V}\cdot\mathbf{x})}{\partial r} - \mathbf{V}\cdot\mathbf{x} \right\}_{r=1} \sin\theta\, d\theta$$

Replacing $\mathbf{V}(r, \theta)$ by its inner expansion

$$\mathbf{V}(r, \theta) = \mathbf{V}_0(r, \theta) + M\mathbf{V}_1(r, \theta) + M^2\mathbf{V}_2(r, \theta) + \cdots$$

we get

$$C_p = 6\pi + \frac{9\pi}{4}M + 4\pi M^2\left(\frac{7}{640} + \frac{\gamma}{20} - \frac{3\gamma^2}{50}\right) + O(M^3)$$

The coefficient C_{m} corresponding to the magnetic force will be computed from the reaction to the Lorentz forces imposed on each fluid particle:

$$C_{\text{m}} = -M^2 \iiint \{\mathbf{V} \times \mathbf{H}_0) \times \mathbf{H}_0\} \cdot \mathbf{x}\, r^2 \sin\theta\, dr\, d\theta\, d\phi$$

the integral being taken over the whole fluid. To compute the coefficient C_m to order $O(M^2)$, it is sufficient to determine the integral to order $O(1)$. To do this, we must express $\mathbf{V}(r, \theta, M)$ in the form

$$\mathbf{V}(r, \theta, M) = \boldsymbol{\alpha}(r, \theta) + \boldsymbol{\beta}(\bar{r}, \theta) + O(M)$$

which is valid for all values of r, i.e., for values such that

$$\lim_{M\to 0} \{\mathbf{V}(r, \theta, M) - \boldsymbol{\alpha}(r, \theta) - \boldsymbol{\beta}(\bar{r}, \theta)\} = 0$$

as well as for r fixed or $\bar{r}$ fixed. We must then have

$$\mathbf{V}_0(r, \theta) = \boldsymbol{\alpha}(r, \theta) + \boldsymbol{\beta}(0, \theta)$$

$$\mathbf{W}_0(\bar{r}, \theta) = \boldsymbol{\alpha}(\infty, \theta) + \boldsymbol{\beta}(\bar{r}, \theta)$$

Taking account of the expressions of $\mathbf{V}_0(r, \theta)$ and $\mathbf{W}_0(\bar{r}, \theta) = \mathbf{x}$, we satisfy all these conditions with

$$\boldsymbol{\alpha}(r, \theta) = \mathbf{V}_0(r, \theta), \qquad \boldsymbol{\beta}(\bar{r}, \theta) = 0.$$

Hence, to compute the coefficient C_m to the order $O(M^2)$, we can replace $\mathbf{V}$ by $\mathbf{V}_0$; we get

$$C_m = \pi M^2 \left(\frac{4\gamma}{5} + \frac{2\gamma^2}{5}\right) + O(M^3)$$

and

$$F = 6\pi\rho_0\nu V_\infty a \left\{1 + \frac{3}{8}M + M^2\left(\frac{7}{960} + \frac{\gamma}{6} + \frac{2\gamma^2}{75}\right)\right\} + O(M^3)$$

In practice, the fluid is a paramagnetic medium, $\mu = 1$, while the spherical shell is either paramagnetic, $\mu' = 1$, or ferromagnetic, $\mu' = \infty$. We have

$$F = 6\pi\rho_0\nu V_\infty\, a \left\{1 + \frac{3M}{8} + \frac{7M^2}{960}\right\} \qquad \text{if } \mu' = 1$$

$$= 6\pi\rho_0\nu V_\infty\, a \left\{1 + \frac{3M}{8} + \frac{321M^2}{1600}\right\} \qquad \text{if } \mu' = \infty$$

In this last case, the drag is larger; in both cases, it is independent of the radius b. (Fig. 5.5.3.a). For intermediate values of μ, the drag

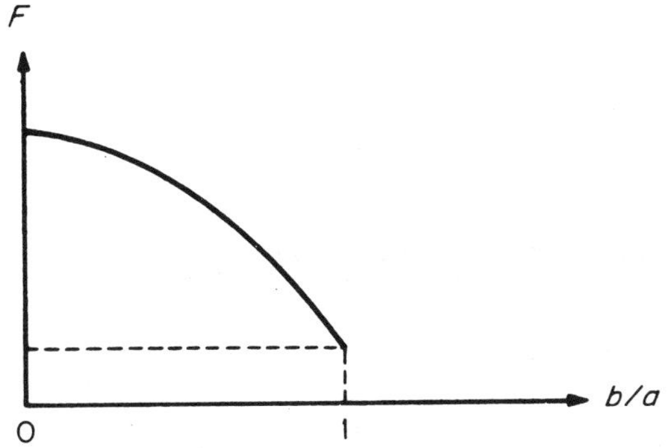

FIG. 5.5.3.a. Drag exerted on a spherical shell. μ, μ', and M constants.

depends on the ratio b/a and is a decreasing function of this ratio; when the sphere is filled, the drag is larger. We have

$$F = 6\pi\rho_0\nu V_\infty a \left\{1 + \frac{3M}{8} + \frac{7M^2}{960}\right\}, \qquad \text{if} \quad b = 0$$

$$= 6\pi\rho_0\nu V_\infty a \left\{1 + \frac{3M}{8} + M^2\left(\frac{7}{960} + \frac{\mu' - \mu}{-(\mu' + 2\mu)} + \frac{2(\mu' - \mu)^2}{75(\mu' + 2\mu)^2}\right)\right\}$$
$$\text{if} \quad b = a$$

It is interesting to note that

$$\lim_{\mu'\to\infty} (\lim_{b\to a} F) \neq \lim_{b\to a} (\lim_{\mu'\to\infty} F)$$

In the case of an infinitesimally thin ferromagnetic shell, the drag is not defined.

CHAPTER 6

FLOW PAST THIN PROFILES

The flow past a profile of small thickness can be treated by means of the linearized equations of magnetofluiddynamics. The case of incompressible flow, treated in Section 6.1, yields results analogous to those of aerodynamics. The case of compressible flow, treated in Section 6.2, yields radically different results; the study of steady flow can be reduced to that of a partial differential equation of the fourth order: two of the characteristics of this equation are always real, while the other two can be either real (purely hyperbolic flows) or imaginary (hyperelliptic flows). While the first two sections deal with steady flows, the third section is concerned with unsteady flows; some of the conclusions reached in this section are quite surprising, so much so that the corresponding results cannot be considered definitive.

6.1. Incompressible Flows

6.1.1. The Linearized Equations

The equations of magnetofluiddynamics in the case of incompressible steady flow can be written in the form

$$\begin{gathered}\rho \mathbf{V} \cdot \nabla \mathbf{V} + \nabla p = \mathbf{J} \times \mu \mathbf{H} \\ \mathbf{J} = \sigma(\mathbf{E} + \mathbf{V} \times \mu \mathbf{H}) \\ \nabla \times \mathbf{H} = \mathbf{J}, \qquad \nabla \times \mathbf{E} = 0 \\ \nabla \cdot \mathbf{V} = 0, \qquad \nabla \cdot \mathbf{H} = 0\end{gathered} \tag{6.1.1.-1}$$

Eliminating **J** and **E**, we find

$$\rho \mathbf{V} \cdot \nabla \mathbf{V} + \nabla p = \mu\{\mathbf{V} \cdot \nabla \mathbf{H} - \nabla(H^2/2)\}$$

$$-\nabla \times (\mathbf{V} \times \mathbf{H}) \equiv \mathbf{V} \cdot \nabla \mathbf{H} - \mathbf{H} \cdot \nabla \mathbf{V} = (1/\sigma\mu)\, \nabla^2 \mathbf{H} \qquad (6.1.1\text{-}2)$$

$$\nabla \cdot \mathbf{V} = \nabla \cdot \mathbf{H} = 0$$

Equations (6.1.1-2) can be linearized in the neighborhood of a constant solution. Taking the case of planar flows, we first assume that the undisturbed velocity and magnetic field are collinear, and later that they are orthogonal.

(1) In the case of aligned flows, we write

$$\mathbf{V} = (U + u, v, 0), \qquad u, v \ll U$$

$$\mathbf{H} = (H_0 + h_x, h_y, 0), \qquad h_x, h_y \ll H_0$$

The first of Eqs. (6.1.1-2) becomes

$$\rho U \frac{\partial \mathbf{v}}{\partial x} + \nabla p = \mu \left\{ H_0 \frac{\partial \mathbf{h}}{\partial x} - H_0\, \nabla h_x \right\}$$

where **v** is the vector (u, v) and **h** the vector (h_x, h_y). This linearized equation gives, when projected on the $0x$ axis,

$$p = p_\infty - \rho U u$$

and, when projected on the $0y$ axis,

$$\Omega = (\mu H_0/\rho U)\, \xi$$

where

$$\Omega = \frac{\partial v}{\partial x} - \frac{\partial u}{\partial y}, \qquad \xi = \frac{\partial h_y}{\partial x} - \frac{\partial h_x}{\partial y} = j_z$$

with j_z denoting the only nonzero component; it is the component along the $0z$ axis of the disturbance of the electric current. The second of Eqs. (6.1.1-2) becomes

$$U \frac{\partial \mathbf{h}}{\partial x} - H_0 \frac{\partial \mathbf{v}}{\partial x} = \frac{1}{\sigma\mu}\, \nabla^2 \mathbf{h}$$

(2) In the case of crossed flows, we write

$$\mathbf{V} = (U + u, v, 0), \qquad u, v \ll U$$

$$\mathbf{H} = (h_x, H_0 + h_y, 0), \qquad h_x, h_y \ll H_0$$

The first of Eqs. (6.1.1-2) becomes

$$\rho H \frac{\partial \mathbf{v}}{\partial x} + \nabla p = \mu \left\{ H_0 \frac{\partial \mathbf{h}}{\partial y} - H_0 \nabla h_y \right\}$$

from which we can deduce

$$\nabla^2(p + \mu H_0 h_y) = 0$$

6.1.2. The Flows with Aligned Fields

The equations determining the disturbed motion are

$$\begin{aligned} \Omega &= \frac{\mu H_0}{4\pi\rho U} \xi \\ U \frac{\partial \mathbf{h}}{\partial x} - H_0 \frac{\partial \mathbf{v}}{\partial x} &= \frac{1}{\sigma\mu} \nabla^2 \mathbf{h} \end{aligned} \tag{6.1.2-1}$$

When the electrical resistivity is negligible, we get

$$\frac{\mathbf{h}}{H_0} - \frac{\mathbf{v}}{U} = \text{function of } \ y = 0 \tag{6.1.2-2}$$

since **h** and **v** tend to zero when x tends to infinity; this proves that the streamlines and the lines of force of the magnetic field are parallel. We also find, from the second equation of (6.1.1-2),

$$\nabla \times (\mathbf{V} \times \mathbf{H}) = 0$$

or $\mathbf{V} \times \mathbf{H} =$ const, since the flow is plane. The first of Eqs. (6.1.2-1) can then be written

$$\left(U^2 - \frac{\mu H_0^{\,2}}{\rho}\right) \Omega = 0 \tag{6.1.2-3}$$

Hence, if the undisturbed velocity is different from the Alfvén speed, the flow is irrotational and we have

$$\Omega = \xi = j_z = 0$$

It follows that the flow is the same as in the case of nonconducting fluid. In the particular case where the undisturbed fluid velocity is equal to Alfvén speed, the disturbances created by the body will propagate as an Alfvén wave, i.e., (with respect to the fluid) with the same velocity as the body. The body then moves in a cavity of the same shape and therefore exerts no force on the fluid.

a. Flow over an Isolated Wavy Wall

The wavy wall is defined by the equation $y = \epsilon \cos \lambda x$, $\lambda\epsilon \ll 1$. The disturbed flow is

$$u - iv = U\epsilon\lambda e^{i\lambda(x+iy)}$$

$$h_x - ih_y = H_0\epsilon\lambda e^{i\lambda(x+iy)}$$

Inside the wall, σ and $\mathbf{V}$ are zero; hence, p is constant and

$$\nabla \times \mathbf{H} = 0, \qquad \nabla \cdot \mathbf{H} = 0$$

Hence, the general solution is

$$h_x + ih_y = f(x + iy)$$

At the boundary, h_y must be continuous; therefore,

$$f(x) = -H_0\epsilon\lambda e^{i\lambda x}$$

so that the internal magnetic field is

$$h_x + ih_y = -H_0\epsilon\lambda e^{i\lambda(x-iy)}$$

The component h_x is then discontinuous across the boundary and we have

$$[h_x(x, 0)] = h_x(x, 0^+) - h_x(x, 0^-) = 2H_0\epsilon\lambda \cos \lambda x$$

The component j_z is infinite, but the quantity

$$\lim_{\epsilon\to 0} \int_{\epsilon^-}^{\epsilon} j_z \, dy, \qquad \epsilon > 0$$

is bounded and has the value $-(1/4\pi)[h_x(x, 0)]$. This limit is called the surface current, J_S. We have

$$J_S = -2H_0\epsilon\lambda \cos \lambda x$$

The pressure of the fluid is $p = p_\infty - \rho U u$, and the pressure on the boundary is

$$p_n(x) = p(x, 0^+) - \mu H_0 J_S$$

since we have

$$\pi = \left(p + \frac{\mu H^2}{2}\right)\mathbf{n} - \frac{\mu H_n \mathbf{H}}{4\pi}$$

$$\delta\pi = (p + \mu\mathbf{H}\cdot\mathbf{h})\, n$$

$$p_n = \left[p + \frac{\mu H_0 h_x}{4\pi}\right] = p(x, 0^+) - \mu H_0 J_S$$

$$p_n(x) = p_\infty - \rho U^2 \epsilon\lambda \cos\lambda x \left\{1 - \frac{2}{m^2}\right\}$$

$$m = \frac{\sqrt{\mu\rho}}{\mu H_0}\, U = \text{Alfvén number}$$

b. The Problem of the Lifting Wing with Zero Thickness

The classical irrotational theory of flow past a body of this type, consists of placing a distribution of vortices on the body, which produces a discontinuity in the u component of the velocity while the v component stays continuous. By Eq. (6.1.2-2), the discontinuity of u is proportional to the discontinuity of h_x at the same point, the latter determining the surface current. Hence, the distribution of lift on the wing is given by

$$\begin{aligned} l_n(x) &= -\rho U\{u(x, -0) - u(x, +0)\} + \mu H_0\{h_x(x, -0) - h_x(x, +0)\} \\ &= 2u(x, +0)\{\rho U - \mu H_0^2\} \\ &= 2\rho U u(x, +0)\{1 - m^{-2}\} \end{aligned}$$

The lift coefficient and the moment coefficient have the values

$$c_l = c_{l_0}(1 - m^{-2}) \qquad \text{and} \qquad c_m = c_{m_0}(1 - m^{-2})$$

where c_{l_0} and c_{m_0} correspond to nonconducting fluids. It is interesting to note that, for a fluid of infinite electrical conductivity, the result will be the same whether the wing is a conductor or is insulated. In both cases, the electric current is the same; it is determined solely by the discontinuity in h_x.

6.1.3. Flow with Crossed Fields

When the electrical resistivity is negligible, the linearized equations (6.1.1-1) have the solution

$$\begin{aligned}
u &= \frac{m}{1+m^2}\{F_1(x-my) - G_1(x+my)\} + \frac{\partial \Phi_1}{\partial x} \\
v &= \frac{1}{1+m^2}\{F_1(x-my) + G_1(x+my)\} + \frac{\partial \Phi_1}{\partial y} \\
h_x &= \frac{m}{1+m^2}\{F_2(x-my) - G_2(x+my)\} + \frac{\partial \Phi_2}{\partial x} \\
h_y &= \frac{1}{1+m^2}\{F_2(x-my) + G_2(x+my)\} + \frac{\partial \Phi_2}{\partial y}
\end{aligned} \tag{6.1.3-1}$$

where Φ_1 and Φ_2 are two harmonic functions, while we have

$$F_2 = -(\rho/\mu)^{1/2} F_1, \qquad G_2 = (\rho/\mu)^{1/2} G_1 \tag{6.1.3-2}$$

and

$$U(\partial\Phi_2/\partial x) = H_0(\partial\Phi_1/\partial y) + \text{const}$$

This constant is zero because of the boundary conditions at infinity. The equation $\mathbf{V} \times \mathbf{H} = \text{const}$ is still valid and gives here $H_0 u + U h_y = 0$, and hence,

$$U(\partial\Phi_2/\partial y) = -H_0(\partial\Phi_1/\partial x)$$

a. The Flow over an Insulated Wavy Wall

The boundary condition $v(x, 0) = -U\lambda\epsilon \sin \lambda\epsilon$ gives

$$\begin{aligned}
G_1 &= G_2 = 0 \\
F_1 &= F_0 \cos \lambda(x-my), \qquad F_0 = \text{const} \\
\Phi_1 - (U/H_0)\,\Phi_2 &= f_1 e^{i\lambda(x+iy)}, \qquad f_1 = \text{const}
\end{aligned}$$

with

$$F_0/(1+m^2) - \lambda f_1 = iU\lambda\epsilon$$

Inside the wall, $\mathbf{V} = 0$ and $\sigma = 0$, we then have

$$h_x - ih_y = -i\lambda f_3 e^{-i\lambda(x+iy)}, \qquad f_3 = \text{const}$$

The boundary condition that h_y is continuous gives

$$\lambda f_3 = -\frac{(\rho/\mu)^{1/2}}{1+m^2} F_0 - i\frac{H_0}{U}\lambda f_1$$

and hence

$$F_0 = -2U\lambda\epsilon\frac{1+m^2}{2i+m+im^2}$$

$$f_1 = U\epsilon\frac{m^2 - im}{2i+m+im^2}$$

We deduce, in particular, the pressure on the boundary

$$p(x, 0) = p_\infty - \rho U^2\lambda\epsilon\frac{m^2 - 2im - 2}{m(m-2i)} e^{i\lambda x}$$

b. The Problem of an Infinitely Thin, Insulated Wing

Assume the boundary conditions to be

$$v(x, 0) = UY'(x) \qquad \text{for} \qquad -b \leqslant x \leqslant b$$

so that the wing profile is $y = Y(x)$. This requires

$$\frac{1}{1+m^2}F_1(x) + \left(\frac{\partial\Phi_1}{\partial y}\right)_{y=+0} = UY'$$

$$\frac{1}{1+m^2}G_1(x) + \left(\frac{\partial\Phi_1}{\partial y}\right)_{y=-0} = UY'$$

We require again the continuity of h_x and h_y on the wing. The first condition leads to

$$\frac{m(\rho/\mu)^{1/2}}{1+m^2}\{F_1(x) - G_1(x)\} = \left(\frac{\partial\Phi_2}{\partial x}\right)_{y=+0} - \left(\frac{\partial\Phi_2}{\partial x}\right)_{y=-0}$$

or

$$m\left(\frac{\rho}{\mu}\right)^{1/2}\left[\frac{\partial\Phi_1}{\partial y}\right]_{y=0} = \left[\frac{\partial\Phi_2}{\partial x}\right]_{y=0}$$

It follows that, on the wing, $\partial\Phi_1/\partial y$ and $\partial\Phi_2/\partial x$ must be continuous. Hence, Φ_1 consists at most of a surface of vortices and Φ_2 of a surface of magnetic sources. We then have $F_1(x) = G_1(x)$, and hence $u(x, +0) = -u(x, -0)$. Since $u(x, y)$ is proportional to $h_y(x, y)$, which is continuous on the wing,

$$u(x, \pm 0) = 0 = h_y(x, 0)$$

Hence, we can connect the rotational and irrotational parts of the flow explicitly,

$$\frac{1}{1+m^2} F_2(x) = -\left(\frac{\partial \Phi_2}{\partial y}\right)_{y=+0}$$

or

$$F_1(x) = -\frac{1+m^2}{m}\left(\frac{\partial \Phi_1}{\partial x}\right)_{y=+0}$$

The boundary condition can therefore be expressed as a function of the unknown potential $\Phi_1(x, y)$ only:

$$\left(\frac{\partial \Phi_1}{\partial x}\right)_{y=+0} - m\left(\frac{\partial \Phi_1}{\partial y}\right)_{y=0} = -mUY'$$

for $-b \leqslant x \leqslant a$. Boundary value problems of this type have been solved by Rott and Cheng in the form

$$\left(\frac{\partial \Phi_1}{\partial x}\right) - i\left(\frac{\partial \Phi_1}{\partial y}\right) = i\left\{\left(\frac{z-b}{z+b}\right)^{\beta} \frac{mU}{\pi(1+m^2)^{1/2}} \int_{-b}^{b} \left(\frac{b+s}{b-s}\right)^{\beta} \frac{Y'(s)\,ds}{s-z}\right\}$$

where $z = x + iy$ and $\tan \pi\beta = m$. Since the meaning of this general formula is not clear, we will show the results in two simple special cases.

(1) In the case of a sinusoidal wing ($Y = \epsilon e^{i\lambda x}$) of infinite chord ($b = \infty$), all the functions are sinusoidal and we obtain

$$\Phi_1(x, y) = -\frac{mi}{m+i} U\epsilon\, e^{i\lambda(x+iy)}, \qquad \text{for} \quad y > 0$$

and hence

$$F_1(x) = -\frac{1+m^2}{m+i} U\lambda\, \epsilon\, e^{i\lambda x}$$

The lift coefficient has the value

$$l(x) = 2\rho U^2 \lambda\, \epsilon\left(1 - \frac{i}{m}\right) e^{i\lambda x}$$

(2) One family of solutions of the equation satisfied by ϕ_1 is given by the Glauert series

$$\left(\frac{\partial \Phi_1}{\partial x}\right)_{y=+0} = B_0 \frac{1-\cos\theta}{\sin\theta} + \sum_{n=1}^{\infty} B_n \sin n\theta$$

$$\left(\frac{\partial \Phi_1}{\partial y}\right)_{y=0} = -B_0 - \sum_{n=1}^{\infty} B_n \cos n\theta$$

with the conditions

$$-mUY'(x) = B_0 \left\{ \frac{1 - \cos \theta}{\sin \theta} + m \right\} + \sum_{n=1}^{\infty} B_n(\sin n\theta + m \cos n\theta)$$

where $x = b \cos \theta$. The wing profile is determined by integration. The lift coefficient has the value

$$l(x) = 2\rho \frac{1 + m^2}{m^2} U \left\{ B_0 \frac{1 - \cos \theta}{\sin \theta} + \sum_{n=1}^{\infty} B_n \sin n\theta \right\}$$

6.2. Compressible Flows

6.2.1. Linearized Equations

The equations of magnetofluiddynamics are satisfied by solutions for which all the parameters of the flow are constant. The corresponding flows are called undisturbed flows; we call **U** and **H** the velocity and magnetic field of an undisturbed flow; other parameters are distinguished by the subscript ∞.

Following the classical method of linearization of the equations, we look for a solution of the form

$$\mathbf{V} = \mathbf{U} + \mathbf{v}, \qquad \mathbf{H} = \mathbf{H}_0 + \mathbf{h}$$

The disturbed quantities are assumed to be small compared to the corresponding undisturbed values. If the axis of the abscissa is taken parallel to **U**, the general equations of magnetohydrodynamics take the form:

Conservation of momentum:

$$\frac{\partial \mathbf{v}}{\partial t} + U \frac{\partial \mathbf{v}}{\partial x} + \frac{1}{\rho_\infty} \nabla p = \frac{\boldsymbol{\xi} \times \mu \mathbf{H}_0}{\rho_\infty} \qquad (6.2.1\text{-}1)$$

with

$$\boldsymbol{\xi} = \nabla \times \mathbf{h}$$

The right-hand side can be modified using the identity

$$\boldsymbol{\xi} \times \mathbf{H}_0 \equiv \mathbf{H}_0 \cdot \nabla \mathbf{h} - \nabla(\mathbf{H}_0 \cdot \mathbf{h})$$

Equation of continuity:

$$\frac{\partial \rho}{\partial t} + U \frac{\partial \rho}{\partial x} + \rho_\infty \nabla \cdot \mathbf{v} = 0 \tag{6.2.1-2}$$

Conservation of energy. The gas is assumed to be polytropic, so that the specific internal energy ϵ takes the value $\epsilon = c_v T$, $c_v = \text{const}$:

$$c_v \left(\frac{\partial T}{\partial t} + U \frac{\partial T}{\partial x} \right) - \frac{p_\infty}{\rho_\infty^2} \left(\frac{\partial \rho}{\partial t} + U \frac{\partial \rho}{\partial x} \right) = 0 \tag{6.2.1-3}$$

Electromagnetic equations:

$$\nabla \times \mathbf{H} = \mathbf{J}, \qquad \nabla \times \mathbf{E} = -\partial \mu \mathbf{H} / \partial t$$

Ohm's law: $\mathbf{J} = \sigma(\mathbf{E} + \mathbf{V} \times \mu \mathbf{H})$

Equation of state: $p = R \rho T$.

Elimination of the electrical field and the current density $\mathbf{J}$ leads to the equation

$$\frac{\partial \mathbf{H}}{\partial t} = \nabla \times (\mathbf{V} \times \mathbf{H}) - \frac{\nabla \times (\nabla \times \mathbf{H})}{\sigma \mu}$$

Taking account of the fact that the divergence of the magnetic field is zero, the last equation can be linearized in the form

$$\frac{\partial \mathbf{h}}{\partial t} + U \frac{\partial \mathbf{h}}{\partial x} = \mathbf{H}_0 \cdot \nabla \mathbf{v} - \mathbf{H}_0 \nabla \cdot \mathbf{v} + \frac{\nabla^2 \mathbf{h}}{\sigma \mu} \tag{6.2.1-4}$$

The linear equations (6.2.1-1)–(6.2.1-4) enable us to determine the unknowns $\mathbf{v}$, $\mathbf{h}$, p, and ρ. The temperature can be eliminated using the equation of state.

6.2.2. The Equation for the Electric Current

Starting from the linearized equations, McCune and Resler were able to form one equation satisfied by the vector $\boldsymbol{\xi}$, equal to the density of the electric current.

The elimination of the velocity between the equation of continuity and the momentum equation leads to Eq. (6.2.2-1); this elimination is carried out by first taking the divergence of both sides of Eq. (6.2.1-1),

$$\nabla^2 p - \frac{1}{Q_\infty^2} \frac{D^2 p}{D t^2} = \mu H_0 \cdot \nabla \times \boldsymbol{\xi} \tag{6.2.2-1}$$

with

$$Q_\infty^2 = \frac{\partial p}{\partial \rho}, \qquad \frac{D}{Dt} = \frac{\partial}{\partial t} + U\frac{\partial}{\partial x}$$

We can then eliminate the pressure by taking the curl of both sides of (6.2.1-1); we get

$$\frac{D\boldsymbol{\Omega}}{Dt} = \frac{\mathbf{H}_0 \cdot \nabla \boldsymbol{\xi}}{\rho_\infty} \tag{6.2.2-2}$$

with $\boldsymbol{\Omega} = \nabla \times \mathbf{V}$.

Finally, we take the curl of both sides of Eq. (6.2.1-4),

$$\frac{D\boldsymbol{\xi}}{Dt} = \mathbf{H}_0 \cdot \nabla\boldsymbol{\Omega} - \frac{\mathbf{H}_0}{\rho_\infty Q_\infty^2} \times \nabla\left(\frac{Dp}{Dt}\right) + \frac{\nabla^2\boldsymbol{\xi}}{\sigma\mu} \tag{6.2.2-3}$$

We can eliminate $\boldsymbol{\Omega}$ between the last two equations and apply the operator $(\nabla^2 - (1/Q_\infty^2)\, D^2/Dt^2)$ to the resultant equation. The elimination of the pressure by means of Eq. (6.2.2-1) leads to the following equation, satisfied by the vector $\boldsymbol{\xi}$ and also the current density $\mathbf{J}$:

$$\frac{D^2\boldsymbol{\xi}}{Dt^2} = \mathbf{H}_0 \cdot \nabla\left[\frac{\mu\mathbf{H}_0 \cdot \nabla\boldsymbol{\xi}}{\rho_\infty}\right] - \frac{\mathbf{H}_0}{\rho_\infty Q_\infty^2} \times \nabla\left(\frac{D^2p}{Dt^2}\right) + \frac{1}{\sigma\mu}\nabla^2\frac{D\boldsymbol{\xi}}{Dt}$$
$$\times\left(\nabla^2 - \frac{1}{Q_\infty^2}\frac{D^2}{Dt^2}\right)\left\{\frac{1}{\sigma\mu}\nabla^2\left(\frac{D\boldsymbol{\xi}}{Dt}\right) + \frac{\mu(\mathbf{H}_0\cdot\nabla)^2\boldsymbol{\xi}}{\rho_\infty} - \frac{D^2\boldsymbol{\xi}}{Dt^2}\right\}$$
$$= \frac{1}{\rho_\infty Q_\infty^2}\frac{D^2}{Dt^2}\{\mu\mathbf{H}_0 \times \nabla(\mathbf{H}_0 \cdot \nabla \times \boldsymbol{\xi})\} \tag{6.2.2-4}$$

The above fifth-order equation is linear and homogeneous.

The equation for the electric current, Eq. (6.2.2-4), can be simplified in certain special cases; we will consider the case of steady plane flow. The flow plane is taken as the x, y plane and the magnetic field is assumed to be parallel to this plane. The electric current is then perpendicular to this plane, $\boldsymbol{\xi} = (0, 0, \xi)$. Equation (6.2.2-4) will have the simplified form

$$\{\beta^2(1 - A_x^2) + (1 - \beta^2)A_y^2\}\frac{\partial^4\xi}{\partial x^4} + \{1 - \beta^2(A_x^2 + A_y^2)\}\frac{\partial^4\xi}{\partial x^2\,\partial y^2}$$
$$- A_y^2\frac{\partial^4\xi}{\partial y^4} - 2A_xA_y\frac{\partial^2}{\partial x\,\partial y}\nabla^2\xi$$
$$= \frac{1}{\sigma\mu U}\left\{\beta^2\frac{\partial^2}{\partial x^2} + \frac{\partial^2}{\partial y^2}\right\}\nabla^2\frac{\partial\xi}{\partial x} \tag{6.2.2-5}$$

where we have written

$$\beta^2 = 1 - \frac{U^2}{Q_\infty{}^2}, \qquad A_x = \frac{H_{0x}}{U}\left(\frac{\mu}{\rho_\infty}\right)^{1/2}, \qquad A_y = \frac{H_{0y}}{U}\left(\frac{\mu}{\rho_\infty}\right)^{1/2}$$

The coefficient A_x^{-1} and A_y^{-1} are called the Alfvén numbers relative to the directions $0x$ and $0y$, respectively.

When Eq. (6.2.2-5) is solved, the electric current $\mathbf{j}$ is known, $\mathbf{j} = \boldsymbol{\xi}$. The other unknowns can then be determined in the following fashion. Equations (6.2.1-1) and (6.2.1-2) are written

$$\rho_\infty U \frac{\partial u}{\partial x} + Q_\infty{}^2 \frac{\partial \rho}{\partial x} = -\mu H_{0y}\xi$$

$$\rho_\infty U \frac{\partial v}{\partial x} + Q_\infty{}^2 \frac{\partial \rho}{\partial y} = \mu H_{0x}\xi \qquad (6.2.2\text{-}6)$$

$$U\frac{\partial \rho}{\partial x} + \rho_\infty\left(\frac{\partial u}{\partial x} + \frac{\partial v}{\partial y}\right) = 0$$

These linear equations with constant coefficients determine the velocity (u, v) and the density ρ. Elimination of the density from the first and third equations gives

$$\beta^2 \frac{\partial u}{\partial x} + \frac{\partial v}{\partial y} = \frac{UH_{0y}}{\rho_\infty Q_\infty{}^2}\,\xi \qquad (6.2.2\text{-}7)$$

To determine the magnetic field, it is sufficient now to use Ohm's law. This states that the electric field is perpendicular to the magnetic field. If this electric field is denoted by $\mathbf{E} + \mathbf{e}$, where $\mathbf{E} = -\mathbf{U} \times \mu\mathbf{H}_0$, the linearized form of Ohm's law can be written

$$\xi = \sigma\{e + uH_{0y} - vH_{0x} + Uh_y\} \qquad (6.2.2\text{-}8)$$

Since $\nabla \times \mathbf{e} = 0$, e is constant. Knowing the component h_y, we then have

$$\partial h_y/\partial x - \partial h_x/\partial y = \xi$$

6.2.3. Flows with Aligned Fields

In the particular case where the undisturbed velocity and magnetic fields are parallel, the reciprocal of the Alfvén number A_y is zero, and in the special case where the electrical resistivity is negligible, the equation determining ξ is reduced to the form

$$\left\{\frac{\beta^2(1 - A_x^2)}{(1 - \beta^2 A_x^2)}\right\} \frac{\partial^2 \xi}{\partial x^2} + \frac{\partial^2 \xi}{\partial y^2} = 0 \tag{6.2.3-1}$$

with the boundary condition at infinity $\xi = 0$.

In the present case, it is easier to introduce the current disturbance function Ψ, the existence of which follows from Eq. (6.2.2-6), with $H_{0y} = 0$:

$$u = \frac{1}{\beta^2} \frac{\partial \Psi}{\partial y}, \qquad v = -\frac{\partial \Psi}{\partial x}$$

The first of Eqs. (6.2.2-6) gives

$$\rho_\infty U u + Q_\infty^2 \rho = 0$$

or, because $(\partial p/\partial \rho) = Q_\infty^2$, $p - p_\infty = -\rho_\infty U u$. The second of Eqs. (6.2.2-6) then gives

$$\frac{\partial v}{\partial x} - \frac{\partial u}{\partial y} = \frac{\mu H_0}{\rho_\infty} \frac{\xi}{U} \tag{6.2.3-2}$$

The linearized form of Ohm's law (6.2.2-8), in which we can assume $e = 0$, then gives

$$h_y = (H_0/U)\, v$$

and since $\nabla \cdot \mathbf{h} = 0$, we obtain, using Eq. (6.2.2-7),

$$h_x = (H_0/U)\, \beta^2 u$$

Expressing $\xi = (\partial h_y/\partial x) - (\partial h_x/\partial y)$ with the aid of u and v, Eq. (6.2.3-2) can be written

$$\frac{\partial v}{\partial x} - \frac{\partial u}{\partial y} = \frac{H_0^2}{\rho_\infty U^2} \left\{\frac{\partial v}{\partial x} - \beta^2 \frac{\partial u}{\partial y}\right\}$$

or

$$\mathscr{B}^2 \frac{\partial^2 \Psi}{\partial x^2} + \frac{\partial^2 \Psi}{\partial y^2} = 0 \tag{6.2.3-3}$$

with

$$\mathscr{B}^2 = \frac{(1 - M_\infty^2)(1 - A_x^2)}{1 - A_x^2 + M_\infty^2 A_x^2}$$

The different regimes that correspond to different values of Alfvén number and Mach number were studied earlier.

The pressure that acts on a body is equal to the discontinuity of the total pressure on the body surface. The total pressure has the value

$$\boldsymbol{\pi} = (p + \tfrac{1}{2}\mu H^2)\,\mathbf{n} - \mu H_n \mathbf{H}$$

The total pressure disturbance

$$\delta\boldsymbol{\pi} = (p + \mu H_0 h_x)\,\mathbf{n}$$

corresponds to the lift coefficient

$$l(x) = \delta\pi(x, 0^-) - \delta\pi(x, 0^+)$$

or

$$l(x) = \rho_\infty U(1 - \beta^2 A_x{}^2)\{u(x, 0^+) - u(x, 0^-)\} \qquad (6.2.3\text{-}4)$$

To write the results explicitly, we have to solve Eq. (6.2.3-3). The nature of the solution depends essentially on the sign of the coefficient $\mathscr{B}^2$. In the plane of coordinates M_∞ and A_x^{-1}, the discussion of the sign of $\mathscr{B}^2$ gives rise to five regions (Fig. 6.2.3.a). Three regions correspond to an elliptic regime and two to a hyperbolic regime.

(*i*) *Elliptic Regimes.* When $\mathscr{B}^2$ is positive, the affine transformation $x = \mathscr{B}x$ transforms Eq. (6.2.3-3) into Laplace's equation. The problem reduces to that of incompressible flow, which was studied in the previous section.

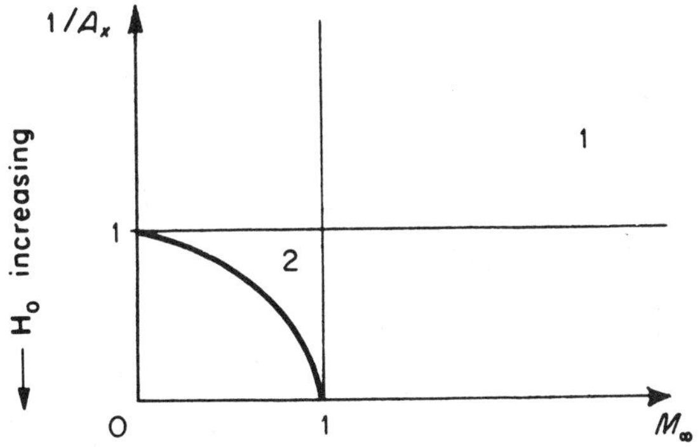

FIG. 6.2.3.a. Five regions resulting from the sign of $\mathscr{B}^2$.

(*ii*) *Hyperbolic Regimes.* When $\mathscr{B}^2$ is negative ($\mathscr{B}^2 = -\mathscr{M}^2$), the general solution of (6.2.3-3) is the following:

$$\Psi(x, y) = F(x - \mathscr{M}y) + G(x + \mathscr{M}y) \qquad (6.2.3\text{-}5)$$

The characteristics cut the x axis at angles arctan $\mathscr{M}^{-1}$.

The hyperbolic regime corresponding to region 1 is analogous to supersonic flow in aerodynamics. In the hyperbolic regime corresponding to region 2, the stable characteristics make an acute angle with the direction of the flow; they are directed upstream. In general, the study of such flows leads to a hyperbolic problem which is not well posed. It follows that the linearization is not valid for subsonic flow of hyperbolic type.

6.2.4. Flows with Crossed Fields

When the velocity and magnetic fields are not aligned in the undisturbed flow, the current disturbance function does not exist. It is then convenient to use the equation satisfied by the current density or ξ. In the special case where the undisturbed velocity and magnetic fields are perpendicular ($A_x = 0$), where $\sigma^{-1} = 0$, this equation becomes

$$(\beta^2 + M_\infty^2 A_y^2)\frac{\partial^4 \xi}{\partial x^4} + (1 - \beta^2 A_y^2)\frac{\partial^4 \xi}{\partial x^2\, \partial y^2} - A_y^2 \frac{\partial^4 \xi}{\partial y^4} = 0 \qquad (6.2.4\text{-}1)$$

This equation can be written in the following form, in which the product of two operators appears:

$$\left\{\frac{\partial^2}{\partial x^2} - R^2 \frac{\partial^2}{\partial y^2}\right\} \times \left\{B\frac{\partial^2}{\partial x^2} + S^2 \frac{\partial^2}{\partial y^2}\right\} \xi = 0 \qquad (6.2.4\text{-}2)$$

with

$$R^2 = \frac{A_y^2\beta^2 - 1 + [4M_\infty^2 A_y^4 + (1 + \beta^2 A_y^2)^2]^{1/2}}{2(\beta^2 + M_\infty^2 A_y^2)}$$

$$S^2 = \frac{A_y^2}{R^2}, \qquad B = 1 - M_\infty^2(1 - A_y^2)$$

The coefficients R^2 and S^2 are positive for all nonzero values of A_y and M_∞. On the other hand, the coefficient B can be either positive or negative; it is always positive if A_y is larger than unity. Hence, the first operator is always hyperbolic, while the second one is hyperbolic or elliptic; it changes type for

$$U^2 = Q_\infty{}^2 + \mu H_0{}^2 = Q_\infty{}^2 + \alpha^2$$

When $B > 0$, the solutions of (6.2.4-2) are called "hyperelliptic"; when $B < 0$, they are called "purely hyperbolic" (Fig. 6.2.4.a).

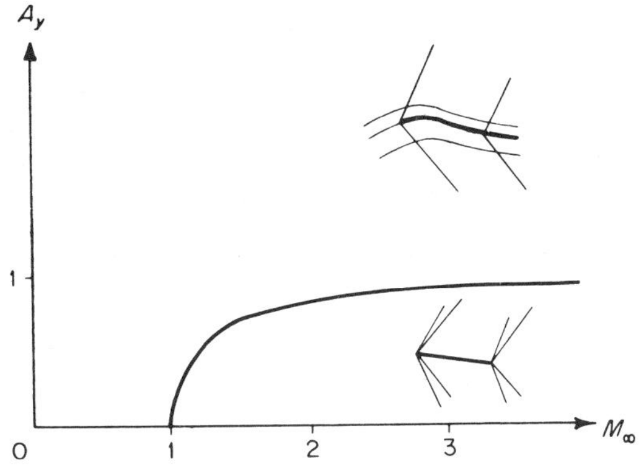

FIG. 6.2.4.a. Flows with crossed fields. Top: hyperelliptic regime; bottom: purely hyperbolic regime.

a. *"Hyperelliptic" Regimes*: $B > 0$

Equation (6.2.4-2) can be written in the form

$$D_h(D_e\xi) = 0 \tag{6.2.4-3}$$

with

$$D_e = B\frac{\partial^2}{\partial x^2} + S^2\frac{\partial^2}{\partial y^2}, \qquad D_h = \frac{\partial^2}{\partial x^2} - R^2\frac{\partial^2}{\partial y^2}$$

In the form (6.2.4-3), the equation for the electric current can be integrated once to give

$$D_e\,\xi = f_1\left(x - \frac{y}{R}\right) + f_2\left(x + \frac{y}{R}\right)$$

A particular solution of the last equation is

$$\xi = \xi_h = F_1\left(x - \frac{y}{R}\right) + F_2\left(x + \frac{y}{R}\right)$$

with

$$\left(B + \frac{S^2}{R^2}\right) F_i''(z) = f_i(z)$$

The general solution, which is also a general solution of Eq. (6.2.4-3), can be written in the form

$$\xi = \xi_{\mathrm{e}} + \xi_{\mathrm{h}}$$

where ξ_{e} and ξ_{h} are arbitrary solutions of the equations

$$D_{\mathrm{e}}\xi_{\mathrm{e}} = 0 \qquad \text{and} \qquad D_{\mathrm{h}}\xi_{\mathrm{h}} = 0$$

We can write the hyperbolic part of the electric current ξ_{h} in the form

$$\xi_{\mathrm{h}} = H_0 \left(1 + \frac{1}{R^2}\right)\left\{F_1'\left(x - \frac{y}{R}\right) + F_2'\left(x + \frac{y}{R}\right)\right\}$$

Knowing that

$$\xi = \frac{\partial h_y}{\partial x} - \frac{\partial h_x}{\partial y} \qquad \text{and} \qquad \frac{\partial h_x}{\partial x} + \frac{\partial h_y}{\partial y} = 0$$

we can easily calculate the hyperbolic parts of the functions h_x and h_y, and hence of u and v,

$$h_x = \frac{H_0}{R}\left\{F_1\left(x - \frac{y}{R}\right) - F_2\left(x + \frac{y}{R}\right)\right\}$$

$$h_y = H_0\left\{F_1\left(x - \frac{y}{R}\right) + F_2\left(x + \frac{y}{R}\right)\right\}$$

$$u = -U\{F_1 + F_2\}$$

$$v = -UR\left\{\beta^2 + M_\infty{}^2 A_y{}^2 \frac{R^2 + 1}{R^2}\right\}\{F_1 - F_2\}$$

$$p = p_\infty + \frac{\rho_\infty U^2}{2} K\{F_1 + F_2\}$$

$$K = 1 - A_y{}^2(M_\infty{}^2 + 1) - [4M_\infty{}^2 A_y{}^4 + (1 + \beta^2 A_y{}^2)^2]^{1/2}$$

To determine the elliptical parts of the same functions, we note that we can write

$$h_x = H_0(\partial\chi/\partial y), \qquad h_y = -H_0(\partial\chi/\partial x)$$

where χ is a certain function of x and y. We then have

$$\xi_{\mathrm{e}} = -H_0\left(\frac{\partial^2\chi}{\partial x^2} + \frac{\partial^2\chi}{\partial y^2}\right)$$

Hence,

$$B\frac{\partial^2\chi}{\partial x^2} + S^2\frac{\partial^2\chi}{\partial y^2} = 0$$

and then

$$u = U\frac{\partial\chi}{\partial x}$$

$$v = U(S^2 - M_\infty{}^2A_y{}^2)\frac{\partial\chi}{\partial y}$$

$$p = p_\infty - \rho_\infty u^2(S^2 - M_\infty{}^2A_y{}^2)\frac{\partial\chi}{\partial x}$$

b. Purely Hyperbolic Regime: $B < 0$

The general solution of Eq. (6.2.4-2) can be written in the form

$$\xi = F_1\left(x - \frac{y}{R}\right) + F_2\left(x + \frac{y}{R}\right) + G_1\left(x - \frac{y}{\Gamma}\right) + G_2\left(x + \frac{y}{\Gamma}\right)$$

where $\Gamma = (-S^2/B)^{1/2}$, and F_1, F_2, G_1, and G_2 are four arbitrary functions. The general expressions for h_x, h_y, u, and v can be deduced very easily.

In the case where the magnetic field, while always in the plane of motion, forms an arbitrary angle with the velocity, we can still use the equation for the electrical current, which, when $\sigma^{-1} = 0$, can be written in the form

$$\{\beta^2(1 - A_x{}^2) + M_\infty{}^2A_y{}^2\}\frac{\partial^4\xi}{\partial x^4} - 2A_xA_y\frac{\partial^2}{\partial x\,\partial y}\nabla^2\xi$$

$$+ \{1 - \beta^2(A_x{}^2 + A_y{}^2)\}\frac{\partial^4\xi}{\partial x^2\,\partial y^2} - A_y{}^2\frac{\partial^4\xi}{\partial y^4} = 0 \quad (6.2.4\text{-}4)$$

The above relation can also be expressed as a product of two operators:

$$\left(\frac{\partial^2}{\partial x^2} + P_0\frac{\partial^2}{\partial x\,\partial y} + Q_0\frac{\partial^2}{\partial y^2}\right) \times \left(\frac{\partial^2}{\partial x^2} + R_0\frac{\partial^2}{\partial x\,\partial y} + S_0\frac{\partial^2}{\partial y^2}\right)\xi = 0$$

$$Q_0S_0 = -A_y{}^2/c_1$$

$$P_0 + R_0 = -2A_xA_y/c_1 = P_0S_0 + R_0Q_0$$

$$Q_0 + P_0R_0 + S_0 = (1 - \beta^2A^2)/c_1$$

$$c_1 = \beta^2 - A_x{}^2 + M_\infty{}^2A^2$$

One of the two numbers Q_0, S_0 is negative. If we assume that Q_0 is negative, then the first operator is hyperbolic; the second will be hyperbolic if $R_0^2 - 4S_0 > 0$, and elliptic if $R_0^2 - 4S_0 < 0$. We have

$$R_0 = -\frac{2A_x A_y}{c_1} \frac{S_0(S_0 - 1)}{(A_y^2/c_1) + S_0^2}$$

$$S_0^3 + \left(\frac{1 - \beta^2 A^2}{c_1} - 3\right) S_0^2 + \frac{A_y^2 + 1 - \beta^2 A^2}{c_1} S_0 - \frac{A_y^2}{c_1} = 0$$

The existence of "hyperelliptic" or doubly hyperbolic flows is hence a general property of steady compressible flows in magnetofluid-dynamics. The degenerate case of aligned fields is the only exception.

6.3. Unsteady Flows

6.3.1. General Equations

In this section, we consider the motion of an incompressible, viscous fluid with electrical conductivity. The equations of motion are written

$$\nabla \cdot \mathbf{V} = 0$$

$$\frac{\partial \mathbf{V}}{\partial t} - \mathbf{V} \times (\nabla \times \mathbf{V}) + \nabla \left(\frac{p}{\rho} + \frac{V^2}{2}\right) = \nu \nabla^2 \mathbf{V} + \frac{(\nabla \times \mathbf{H}) \times \mu \mathbf{H}}{\rho} \tag{6.3.1-1}$$

$$\frac{\partial \mathbf{H}}{\partial t} - \nabla \times (\mathbf{V} \times \mathbf{H}) = \frac{\nabla^2 \mathbf{H}}{\sigma \mu}$$

One interesting special case is that of plane steady flow of a fluid of unit permeability, negligible viscosity, and infinite electrical conductivity. If we further assume that the magnetic field and the velocity are collinear at infinity, the equations simplify considerably. We use an orthogonal frame of reference $0xyz$ such that $\mathbf{V}$ and $\mathbf{H}$ are parallel to the plane $z = 0$. We can then write

$$u = \partial \Psi / \partial y, \qquad v = -\partial \Psi / \partial x$$

where $\mathbf{V} = (u, v, 0)$ and Ψ is the stream function. Knowing that $\sigma^{-1} = 0$, we have $\mathbf{V} \times \mathbf{H} = A\mathbf{k}$, where $\mathbf{k}$ is the unit vector parallel to $0z$ and A is a scalar constant; this constant is zero because $\mathbf{V}$ and $\mathbf{H}$ are aligned at infinity. Hence $\mathbf{H} = \alpha \mathbf{V}$, where α is a scalar function of position only, and, further, since $\nabla \cdot \mathbf{H} = 0$, it is a function of Ψ.

Substituting in the equations of motion and puting $\nu = 0$, we get

$$\mathbf{V} \times (\nabla \times \mathbf{V}) = \nabla \left(\frac{p}{\rho} + \frac{V^2}{2}\right) - \frac{\alpha(\Psi)}{\rho} \{\alpha(\Psi)\, \nabla \times \mathbf{V} - \alpha'(\Psi)\, \mathbf{V} \times \nabla\Psi\} \times \mathbf{V}$$

or

$$\left\{\left(1 - \frac{\alpha^2}{\rho}\right) \nabla^2\Psi - \frac{\alpha\alpha'}{\rho} V^2\right\} \nabla\Psi = \nabla \left(\frac{p}{\rho} + \frac{V^2}{2}\right)$$

so that

$$\left(1 - \frac{\alpha^2}{\rho}\right) \nabla^2\Psi = \frac{\alpha\alpha'}{\rho} (\nabla\Psi)^2 + \beta(\Psi)$$

where $\beta(\Psi)$ and $\alpha(\Psi)$ are functions that must be determined before solving the equation for Ψ.

Assuming that, at infinity upstream, the velocity is uniform with components $(U, 0, 0)$ and that the magnetic field is also uniform there with components $(H_\infty, 0, 0)$, we have

$$\alpha(\Psi) = H_\infty/U, \qquad \beta(\Psi) = 0$$

because, when $x \to \infty$, $\Psi \to Uy$. Hence, $\nabla^2\Psi = 0$ on all streamlines coming from infinity. If we are sure that the flow does not contain any closed streamlines and that no streamlines start and finish at infinity downstream, then we can conclude that the flow is the same as if there were no magnetic field and that lines of force coincide with the streamlines. Moreover, the boundary conditions on a solid paramagnetic insulator are satisfied, since the magnetic field is tangential to the surface.

However, in a steady flow started from rest, we cannot be absolutely sure that the function $\alpha(\Psi)$ is a constant and that there are no streamlines starting from infinity downstream. If the fluid is set in motion with constant velocity $(U, 0, 0)$ at infinity and flows past a fixed body, the velocity field at the time $t = 0$ is certainly irrotational. If, moreover, a constant magnetic field $(H_\infty, 0, 0)$ is applied at infinity, the magnetic field is still uniform at $t = 0$. We then have $\nabla \times (\mathbf{V} \times \mathbf{H}) = 0$ at $t = 0$, so that the magnetic field must start to change. A part of this change corresponds to convection of the magnetic field by the fluid, another part is due to convection by Alfvén waves in such a way that the new field satisfies the boundary conditions on the body. These waves start at the body and propagate with velocity $|\mathbf{H}|/\sqrt{\rho}$†. The waves propagating downstream reach

† In what follows, the value $\mu = 1$ is used.

infinity where, unless their intensity goes to zero, they modify the flow and consequently the values of α and β. We can see that they produce a reversed flow behind the body. Similar considerations are also valid for the upstream region. If we have $U < H_\infty/\sqrt{\rho}$, Alfvén waves reach upstream infinity and change the values of α and β causing a rotational motion. The peculiarities relative to the case $U > H_\infty/\sqrt{\rho}$ will be apparent in examples; but before studying them, we must formulate the boundary conditions precisely.

We consider the problem of a body in an inviscid, perfectly conducting fluid and we assume that, at time $t = 0$, the fluid at infinity moves with velocity U in the direction of the positive $0x$ axis. At the same instant or at an earlier time, a magnetic field parallel to this velocity and of magnitude H_∞ is imposed on the flow. The boundary conditions to be stated are as follows: First, at time $t = 0$, the velocity field is the irrotational field which would arise in the absence of magnetic field. At this time, the magnetic field is still uniform and has a magnitude H_∞. Second, at large distances and for all finite times, $\mathbf{V} \to (U, 0, 0)$, $\mathbf{H} \to (H_\infty, 0, 0)$, because all disturbances originate at the body.

To write down the boundary conditions on the body, some precautions must be taken. We assume that the body is made of paramagnetic, nonconducting material, which means that $\sigma = 0$ and that no magnetic source exists inside the body. Two boundary conditions are that the normal components of velocity and magnetic field are zero on the body; the tangential component of the magnetic field is therefore arbitrary. In the general case, we cannot assume that the normal component of magnetic field is zero and we will get conditions to be applied to the tangential component.

In the case where the boundary of a perfectly conducting fluid is a free surface, the tangential component of the magnetic field must be continuous. One initial discontinuity is immediately propagated in the fluid along an Alfvén wave. This result is established by considering an inviscid fluid with very high but finite electrical conductivity and considering the limiting process $\sigma \to \infty$. In the case of viscous fluid, the effects of viscosity on the magnetic field can be studied with the following example of steady flow.

A fluid with electrical conductivity σ and kinematic viscosity ν is in contact with an insulated solid along the plane $x = 0$. Inside the solid, occupying the region $x < 0$, the magnetic field has components $(H_0, H_-, 0)$; in the fluid, the magnetic field has the components

$(H_0, h, 0)$ and velocity $(0, v, 0)$. In the fluid far from the solid boundary, $v \to U$ and $h \to H_+$. The motion is one-dimensional, so that v and h are functions of x only, all other quantities being constant. The equations of motion reduce to

$$-H_0 \frac{\partial v}{\partial x} = \frac{1}{\sigma} \frac{\partial^2 h}{\partial x^2}, \qquad \frac{-H_0}{\rho} \frac{\partial h}{\partial x} = \nu \frac{\partial^2 v}{\partial x^2}$$

The appropriate boundary conditions are

$$v \to U, \quad h \to H_+ \qquad \text{for} \quad x \to +\infty$$

$$v = 0, \quad h = H_- \qquad \text{for} \quad x = 0$$

because in a fluid with finite conductivity, all components of the magnetic field are continuous on the boundary. The solution of this problem is

$$h = H_+ + (H_- - H_+) \exp \left\{ -H_0 \times \left(\frac{\sigma}{\rho\nu} \right)^{1/2} \right\}$$

$$v = \frac{H_+ - H_-}{(\rho\gamma\sigma)^{1/2}} \left\{ 1 - \exp \left[-H_0 \times \left(\frac{\sigma}{\rho\nu} \right)^{1/2} \right] \right\}$$

where $U = (H_+ - H_-)/(\rho\nu\sigma)^{1/2}$ cannot be arbitrary. If we now take the case of a perfectly conducting, inviscid fluid ($\nu \to 0$, $\sigma \to \infty$), we find

$$h = H_-, \qquad v = 0 \qquad \text{for} \quad x < 0$$

$$h = H_+, \qquad v = \frac{H_+ - H_-}{(\rho\nu\sigma)^{1/2}} \qquad \text{for} \quad x > 0$$

A magnetodynamic boundary layer of a particularly simple type appears on the plane $x = 0$. Hence, in this particular problem, it is not necessary that the tangential component of the velocity be zero on the obstacle, nor that the tangential component of the magnetic field be continuous. However, we must have

$$[\mathbf{H}]_t = (\rho\nu\sigma)^{1/2}[\mathbf{V}]_t$$

where $[\mathbf{V}]_t$ represents the discontinuity of the tangential component. The value of the product $\nu\sigma$ is indeterminate for an inviscid, perfectly conducting fluid, but can be measured for a fluid of really high conductivity and low viscosity. In the problems treated at present (terrestrial or cosmic), this product is negligible and we can use the boundary condition $[\mathbf{H}]_t = 0$.

Even though this condition was derived in a special case, there is no difficulty in showing that it is generally true.

6.3.2. Unsteady Perturbations

In this section, we will consider perturbations caused by a thin profile placed on the x axis, assuming that, at time $t < 0$, the magnetic field has the components $(H_\infty, 0, 0)$ and that, at time $t = 0$, the fluid is set in motion with a velocity at infinity $(U, 0, 0)$. Since the body is thin, we can write

$$\begin{aligned} \mathbf{V} &= (U + u, v, 0) \\ \mathbf{H} &= (H_\infty + h_x, h_y, 0) \end{aligned} \tag{6.3.2-1}$$

where u, v, h_x, and h_y are small. Neglecting the squares and products of these quantities and putting ν and σ^{-1} equal to zero, the equations of motion become

$$\begin{aligned} &\frac{\partial h_x}{\partial x} + \frac{\partial h_y}{\partial y} = 0, \qquad \frac{\partial u}{\partial x} + \frac{\partial v}{\partial y} = 0 \\ &\frac{\partial h_x}{\partial t} + U\frac{\partial h_x}{\partial x} = H_\infty \frac{\partial u}{\partial x}, \qquad \frac{\partial h_y}{\partial t} + U\frac{\partial h_y}{\partial x} = H_\infty \frac{\partial v}{\partial x} \\ &\frac{\partial u}{\partial t} + U\frac{\partial u}{\partial x} = \frac{\partial P}{\partial x} + \frac{H_\infty}{\rho}\frac{\partial h_x}{\partial x} \\ &\frac{\partial v}{\partial t} + U\frac{\partial v}{\partial x} = \frac{\partial P}{\partial y} + \frac{H_\infty}{\rho}\frac{\partial h_y}{\partial x} \end{aligned} \tag{6.3.2-2}$$

with

$$P = -\frac{p}{\rho} - \frac{H^2}{2\rho} + \text{const}$$

These equations can be separated into two groups. The first, due directly to P, corresponds to an irrotational flow:

$$h_x^{(1)} = H_\infty \frac{\partial^2 \phi}{\partial x^2}, \qquad h_y^{(1)} = H_\infty \frac{\partial^2 \phi}{\partial x\, \partial y}$$

where

$$\frac{\partial^2 \phi}{\partial x^2} + \frac{\partial^2 \phi}{\partial y^2} = 0$$

It follows that

$$u^{(1)} = \frac{\partial^2 \phi}{\partial t\, \partial x} + U \frac{\partial^2 \phi}{\partial x^2}, \qquad v^{(1)} = \frac{\partial^2 \phi}{\partial t\, \partial y} + U \frac{\partial^2 \phi}{\partial x\, \partial y}$$

$$P = \frac{\partial^2 \Phi}{\partial t^2} + 2U \frac{\partial^2 \Phi}{\partial t\, \partial x} + \left(U^2 - \frac{H_\infty{}^2}{\rho}\right) \frac{\partial^2 \Phi}{\partial x^2}$$

A second group, independent of P, corresponds to the hyperbolic equations

$$\left(\frac{\partial}{\partial t} + U \frac{\partial}{\partial x}\right)^2 u^{(2)} = \frac{H_\infty{}^2}{\rho} \frac{\partial^2 u^{(2)}}{\partial x^2}, \qquad \frac{H_\infty}{\rho} \frac{\partial h_x^{(2)}}{\partial x} = \frac{\partial u^{(2)}}{\partial t} + U \frac{\partial u^{(2)}}{\partial x}$$

with analogous equations for $v^{(2)}$ and $h_y^{(2)}$. The two groups of functions are related by Eqs. (6.3.2-1). These solutions correspond to Alfvén waves, and, from the small-disturbance theory, the two solutions can be added.

The solution for $u^{(2)}$ is

$$\frac{u^{(2)}}{U} = f\left\{t - \frac{mx}{(1+m)\,U}, y\right\} + g\left\{t + \frac{mx}{(1-m)\,U}, y\right\}$$

where the functions f and g are arbitrary and where $H_\infty m = U\sqrt{\rho}$, so that m is the ratio of the velocity of the undisturbed fluid to the Alfvén speed. All disturbances originate at the body, none can come from infinity. Hence, for $m < 1$,

$$u^{(2)} = Uf\left\{t - \frac{mx}{(1+m)\,U}, y\right\}, \qquad h_x^{(2)} = -mH_\infty f \qquad \text{if} \quad x > 0$$

$$u^{(2)} = Ug\left\{t + \frac{mx}{(1-m)\,U}, y\right\}, \qquad h_x^{(2)} = mH_\infty g \qquad \text{if} \quad x < 0$$

and for $m > 1$,

$$u^{(2)} = u(f+g), \qquad h_x^{(2)} = mH_\infty(f-g) \qquad \text{if} \quad x > 0$$

$$u^{(2)} = 0, \qquad h_x^{(2)} = 0 \qquad \text{if} \quad x < 0$$

When the time is increased indefinitely, these functions tend to finite limits if, as we assume, the motion becomes steady. Writing $f(\infty, y) = f(y)$, $g(\infty, y) = g(y)$, we get, for $m < 1$,

$$u^{(2)} = Uf(y), \qquad h_x^{(2)} = -mH_\infty f(y), \qquad v^{(2)} = h_y^{(2)} = 0 \qquad \text{if} \quad x > 0$$

$$u^{(2)} = Ug(y), \qquad h_x^{(2)} = mH_\infty g(y), \qquad v^{(2)} = h_y^{(2)} = 0 \qquad \text{if} \quad x < 0$$

and for $m > 1$,

$$u^{(2)} = U\{f(y) + g(y)\}, \qquad v^{(2)} = 0,$$

$$h_x^{(2)} = mH_\infty\{g(y) - f(y)\}, \qquad h_y^{(2)} = 0 \qquad \text{if} \quad x > 0$$

$$u^{(2)} = v^{(2)} = h_x^{(2)} = h_y^{(2)} = 0 \qquad \text{if} \quad x < 0$$

Adding the two series of functions with superscript $^{(1)}$ and $^{(2)}$, we get the complete solution, assuming that the disturbances stay small and that the limiting motion is steady.

Inside the body, which is assumed to be insulated and paramagnetic, the magnetic field is written in the form

$$h_x = H_\infty \frac{\partial^2 \Psi}{\partial x^2}, \qquad h_y = H_\infty \frac{\partial^2 \Psi}{\partial x\, \partial y}$$

where the function Ψ is analytic and harmonic inside the obstacle.

It is now easy to apply these results to the study of the flow past a thin, convex profile symmetric about the x axis, the shape of which is given by

$$|\, y \,| = \epsilon S(x), \qquad -b \leqslant x \leqslant a$$

where ϵ is small and $S(x)$ is a function of x of the first order which has a bounded derivative equal to zero for $x = -b$ and $x = a$. Then the boundary conditions to be satisfied imply that, on the wing,

$$v = \epsilon US'(x)\, y/|\, y \,|$$

and hence

$$\frac{\partial \Phi}{\partial x} = \frac{\epsilon}{2\pi} \int_{-b}^{a} S'(\xi) \log\{(x - \xi)^2 + y\}\, d\xi$$

On the other hand, to ensure continuity of **V** and **H** in the fluid, the plane $x = 0$ must cross the wing at the section of maximum area; therefore,

$$f(y) = g(y) = 0, \qquad \text{if} \qquad |\, y \,| > \epsilon S(0)$$

Since $Uh_y = H_\infty v$ outside the wing,

$$h_y = \epsilon H_\infty S'(x) \operatorname{sign}[y]$$

on the wing. It follows that on the wing, to the first approximation, we have

$$h_y = H_\infty y S'(x)/S(x)$$

Using the equation of continuity, it follows that

$$h_x = H_\infty \log S(x) + \text{const}$$

inside the wing for the first approximation of ϵ. The additive constant is undetermined in thin-wing theory. Hence, even though h_y is small inside the wing, h_x is of the same order as H_∞ and the field is seriously disturbed.

Since we have on the wing

$$\frac{\partial^2 \phi}{\partial x^2} = \frac{\epsilon}{\pi} \int_{-b}^{a} \frac{S'(\xi)}{x - \xi}\, d\xi$$

it follows that

$$mg(y) = -\frac{\epsilon}{\pi} \int_{-b}^{a} \frac{S'(\xi)\, d\xi}{x - \xi} + \log S(x) + \text{const}$$

for $-b < x < a$, and $y = \epsilon S(x)$; we have a similar formula for $f(y)$. Hence, if $m < 1$, the magnetic field and the velocity are modified considerably between the planes $y = \pm\epsilon S(0)$, and since this region is the most important part of the flow, the hypothesis of small disturbances does not hold. When $m > 1$, the situation is still contradictory and no solution is possible.

The study of the flow past a body started from rest then shows that no such flow can exist in which disturbances remain small. The reason for this is that, ahead of the obstacle, according to small-disturbance theory, Alfvén waves cannot be produced. Then, on the portion of the boundary where we have $x < 0$, the continuity of h_x and h_y implies that, inside the body, the magnetic field is given by an analytic continuation of Φ. But at the time $t = 0$, the function Φ and the magnetic field are given by distribution of sources on $0x$ inside the wing. Since the wing is paramagnetic, this result is contradictory. In the case where $m > 1$, we note that the difficulties of the theory are due entirely to the linearization of the equations. When the exact equations are used, the Alfvén waves can propagate from the wing into the fluid.

The general conclusion is the following. If a nonconducting thin wing is set in motion with uniform velocity $U \neq 0$ parallel to itself in a perfectly conducting fluid, then the disturbances created by the wing do not stay small. There is, however, an exception; if $U \ll 1$, it is possible to develop a satisfactory linearized theory because the velocity itself, and not merely the disturbance velocity, remains very small. The study of this case is presented in the next section.

6.3.3. Slow Motions

The equations of motion are still the same, but we neglect terms containing U as a factor. We assume, until found otherwise, that the disturbances remain very small; note that this does not mean that the wing is necessarily thin and that the theory, if it proves successful, could be applied to any body. The general solution of the equations of motion is then:

a. For $x > 0$,

$$h_x = H_\infty \frac{\partial^2 \Phi}{\partial x^2} - U\rho^{1/2} f\left(t - \frac{x}{V}, y\right)$$
$$u = \frac{\partial^2 \Phi}{\partial t\, \partial x} + Uf\left(t - \frac{x}{V}, y\right) \tag{6.3.3-1}$$

b. For $x < 0$,

$$h_x = H_\infty \frac{\partial^2 \Phi}{\partial x^2} + U\rho^{1/2} g\left(t + \frac{x}{V}, y\right)$$
$$u = \frac{\partial^2 \Phi}{\partial t\, \partial x} + Ug\left(t + \frac{x}{V}, y\right) \tag{6.3.3-2}$$

where Φ is a harmonic function of the same order as U, while $V = H_\infty/\rho^{1/2}$. We write the values of v and h_y in the same way. In particular, if the flow is ultimately steady, then, for $x > 0$,

$$u = Uf(y), \qquad v = 0, \qquad h_x = H_\infty \frac{\partial^2 \Phi}{\partial x^2} - u\rho^{1/2} f(y), \qquad h_y = H_\infty \frac{\partial^2 \Phi}{\partial x\, \partial y}$$

and for $x < 0$,

$$u = Ug(y), \qquad v = 0, \qquad h_x = H_\infty \frac{\partial^2 \Phi}{\partial x^2} + u\rho^{1/2} g(y), \qquad h_y = H_\infty \frac{\partial^2 \Phi}{\partial x\, \partial y}$$

On the other hand,

$$P = \text{const} - \frac{H_\infty{}^2}{\rho} \frac{\partial^2 \Phi}{\partial x^2}$$

Let us assume now that the obstacle is convex and let S be its boundary. Then the plane $x = 0$ will cut the wing along the section of maximum area; if, moreover, the body is between the two planes $y = \pm C$, it follows that

$$g(y) = f(y) = 1 \qquad \text{for} \quad |\, y \,| < C$$

$$g(y) = f(y) = 0 \qquad \text{for} \quad |\, y \,| > C$$

such that, on S, the condition relative to the velocity is satisfied. We deduce that the fluid is at rest in the region $|\, y \,| < C$, and has the undisturbed velocity U in the region $|\, y \,| > C$. The magnetic field is given by

$$h_x = H_\infty \frac{\partial^2 \Psi}{\partial x^2}, \qquad h_y = H_\infty \frac{\partial^2 \Psi}{\partial x\, \partial y}$$

inside S, where Ψ is a regular harmonic function. Since all of the components of $\mathbf{H}$ are continuous on S, we have that, on S,

$$\frac{\partial^2 \Psi}{\partial x\, \partial y} = \frac{\partial^2 \Phi}{\partial x\, \partial y}$$

on S and $x > 0$,

$$\frac{\partial^2 \Psi}{\partial x^2} = \frac{\partial^2 \Phi}{\partial x^2} - \frac{U}{V}$$

and on S and $x < 0$,

$$\frac{\partial^2 \Psi}{\partial x^2} = \frac{\partial^2 \Phi}{\partial x^2} + \frac{U}{V}$$

The expressions for Ψ and Φ depend on the shape of S. For example, let us consider the case of a circular cylinder of radius $r = c$ (with $x = r \cos \theta$, $y = r \sin \theta$). Writing

$$\frac{\partial \Phi}{\partial x} = -A_0 \log r + \sum_{n=1}^{\infty} A_n \left(\frac{c}{r}\right)^n \frac{\cos n\theta}{n}$$

$$\frac{\partial \Psi}{\partial x} = B_0 + \sum_{n=1}^{\infty} B_n \left(\frac{r}{c}\right)^n \frac{\cos n\theta}{n}$$

we obtain at first $B_n = A_{n-2}$, $n \geqslant 2$. Then

$$B_1 = 0, \qquad A_{2s+1} = 0, \qquad A_{2s} = \frac{(-)^{s+1}\, 2U}{\pi(2s+1)V}, \qquad s = 0, 1, 2, \ldots$$

and hence

$$\frac{\partial \Phi}{\partial x} = \frac{U}{2\pi V}\Big\{4c - 2x \arctan \frac{2cx}{r^2 - c^2} + (y - c)\log[x^2 + (y - c)^2] + (y + c)\log[x^2 + (y + c)^2]\Big\}$$

and $\partial \Psi / \partial x$ differs from $\partial \Phi / \partial x$ only by a constant. Hence, we find that the magnetic field has a logarithmic singularity at the points $x = 0$, $y = \pm C$, which means that at these points the hypothesis of small disturbances is not valid. However, as U is small, the region in which the small-disturbance hypothesis does not hold is exponentially small and it is probable that the error in these regions is very small.

At large distances from the cylinder, the magnetic field is the sum of

(1) A field parallel to $0x$ and of magnitude $H_\infty - U(\rho)^{1/2}$ sign$[x]$ for $|y| < c$, and H_∞ for $|y| > c$.

(2) A field produced by a simple pole at the origin, of intensity $(2U/\pi)(\rho)^{1/2}$.

The computation of the pressure gives, when $t \to \infty$,

$$p \to p_0 - \rho/^{1/2}\, UH_\infty \operatorname{sign}[x] \qquad \text{for} \quad |y| < c$$

$$p \to p_0 \qquad \text{for} \quad |y| > c$$

where p_0 is a constant. Hence, for a circular cylinder, the force per unit thickness has the value $2c[(\rho/\pi)\, UH_\infty]^{1/2}$. This value is valid for any convex profile of thickness $2c$.

The general conclusion of this section is that the study of steady motion of a nonconducting body in an inviscid, perfectly conducting fluid, in the presence of a magnetic field parallel to the velocity, is an extremely complicated problem. If we consider the steady case alone, the problem is undetermined and depends on two arbitrary functions of the stream function. If the conditions at infinity (upstream and downstream) are uniform, these functions can be determined. But, in general, such a simple situation does not occur very often, and the arbitrary function can only be determined in the manner used for steady flow.

CONCLUSION

In concluding this book, we wish to emphasize that the questions treated do not give a complete idea of magnetofluiddynamics. We have chosen those problems which would interest workers in the fields of aeronautical and space sciences most directly. Even with this limitation, we have not been able to present a complete picture—one ought to consider the dissipation of a discontinuity, problems of shock tubes, conical flows, axisymmetric flows, flows of mixtures, and so forth.

The scientific interest of magnetohydrodynamics is not in question, and the results obtained show that we are able to discuss most of the problems, however difficult they may be. The more disputable points, perhaps, and those where a special effort seems to be needed, concern the hypotheses and assumptions used. Some of the surprising results and contradictions that we have encountered could be cleared up by a more precise presentation or modification of the hypotheses used to solve the problems of magnetofluiddynamics. The boundary conditions to be used are still an open question. The few notes on the airfoil theory show very well the difficulties encountered. When a solution is found theoretically, we have to decide whether or not it corresponds to reality.

An important point therefore concerns laboratory experiments designed to reproduce the phenomena of magnetofluiddynamics. The difficulty is to produce strong enough magnetic fields and to find fluids with a sufficiently large conductivity for the phenomena studied theoretically to be measured. While mercury is an ideal fluid as far as electrical conductivity is concerned, the difficulty is to find a gas with such a property. One can hope, however, that most of the sub-Alfvénic phenomena—upstream wakes, subsonic hyperbolic flows, and supersonic elliptic flows—will be explored experimentally one day.

In the problems that interest us most (flight of rockets, motion of satellites), conditions are different. The atmosphere is a weak conductor because it is ionized by radioactive substances and cosmic rays; however, the electrical conductivity stays small (0.1 mho/cm for the case of a satellite, for example). The magnetic field of the earth is also very small; at large distances, its distribution is that of a dipole of magnetic moment 8×10^{25} C.G.S. placed at the earth's center, the axis of which pierces the earth at the magnetic poles. The use of magnetofluiddynamic effects would necessitate the artificial increase of electrical conductivity in the neighborhood of the rockets and satellites, for example, by taking molecules from the metal covering of the rockets; it would also necessitate the creation of artificial magnetic fields. In both cases, the weight of the equipment would create difficult problems for the engineer.

It would, however, be very unwise to state that magnetofluiddynamics offers little interest for aeronautics. The past has taught us that any science can develop in surprising directions, and magnetofluiddynamics can perhaps be the means for new progress in aeronautical navigation, and even play a fundamental role in astronautics.

REFERENCES

General

Alfvén, H. (1950). "Cosmical Electrodynamics." Oxford Univ. Press, London and New York.

Bershader, D. (1959). "The Magnetodynamics of Conducting Fluids." Stanford Univ. Press, Stanford, California.

Cabannes, H. (1962). Problèmes de magnétodynamique des fluides. *Proc. 3rd Intern. Congr. Aeron. Sci., Stockholm, 1962.* Mac Millan, New York.

Cambel, A. (1963). "Plasms Physics and Magnetofluid-Mechanics." McGraw-Hill, New York.

Cowling, T. G. (1957). "Magnetohydrodynamics." Wiley (Interscience), New York.

Ferraro, V., and Plumpton, C. (1961). "An Introduction to Magneto-Fluid Mechanics." Oxford Univ. Press, London and New York.

Germain, P. (1959). Introduction à l'étude de l'aéromagnétodynamique. *Cahiers Phys.* **103**, 98–128.

Kulikovskiy, A., and Lyubimov, G. (1962). "La Magnétodynamique des Fluides" (in Russian). Editions d'État, Moscow.

Landau, L., and Lifshitz, E. (1960). "Electrodynamics of Continuous Media." Pergamon, New York.

Landshoff, R. (1960). "Magnetohydrodynamics," 3rd ed. Stanford Univ. Press, Stanford, California.

Mannal, C., and Mather, N. (1962). "Engineering Aspects of Magnetohydrodynamics." Columbia Univ. Press, New York and London.

Napolitano, L., and Contursi, G. (1962). "Magneto-Fluid Dynamics." Pergamon, New York.

Resler, E., and Sears, W. (1958). The prospects for magneto-aerodynamics. *J. Aeron. Sci.* **25**, 235–245.

Shercliff, J. (1965). A Textbook of Magnetohydrodynamics," Pergamon Press, New York and London.

Soubbaramayer, (1967). Sur les chocs dans un milieu magnétodynamique réactif avec application au problème du piston. Thesis, Univ. of Paris, Paris, France

Sutton, G., and Sherman, A. (1965). "Engineering Magnetohydrodynamics." McGraw-Hill, New York.

Chapter I

Chu, B. (1959). Thermodynamics of electrically conducting fluids. *Phys. Fluids* **2**, 473–484.

Lundquist, S. (1952). Studies in magnetohydrodynamics. *Arkiv Fysik* **5**, 347.

Chapter II

Bohachevsky, I. (1962). Simple waves and shocks, in magnetohydrodynamics. *Phys. Fluids* **5**, 1456–1467.

Courant, R., and Friedrichs, K. (1948). "Supersonic Flow and Shock Waves." Wiley (Interscience), New York.

Cumberbatch, E. (1962). Magnetohydrodynamic Mach cones. *J. Aerospace Sci.* **29**, 1476–1479.

Friedrichs, K. (1955a). Non linear wave motion in magnetohydrodynamics. Rept. No. 1845. Los Alamos Develop. Center, Los Alamos, New Mexico.

Friedrichs, K. (1955b). Nichtlineare differentzielgleichungen. Lecture notes. Göttingen Univ., Göttingen.

Hoffman, F., and Teller, E. (1950). Magnetohydrodynamic shocks. *Phys. Rev.* **80**, 692–703.

Jeffrey, A., and Taniuti, T. (1964). "Non Linear Waves Propagation with Applications to Physics and Magnetohydrodynamics." Academic Press, New York.

Lax, P. (1957). Hyperbolic systems of conservation laws. *Comm. Pure Appl. Math.* **10**, 537–566.

Lighthill, M. (1960). Studies on magnetohydrodynamic waves and other anisotropic waves motions. *Phil. Trans. Roy. Soc. London Ser. A* **252**, 397–430.

Ludford, G. (1959). The propagation of small disturbances in hydromagnetics. *J. Fluid Mech.* **5**, 387–400.

Polovin, R. (1961). Contribution to the theory of simple magnetohydrodynamic waves. *Soviet Phys. JETP (English Transl.)* **12**, 326–330.

Resler, E. (1960). Some remarks on hydromagnetic waves for finite conductivity. *Rev. Mod. Phys.* **32**, 866–867.

Rott, N. (1959). A simple construction for the determination of the magnetohydrodynamic wave speed in a compressible conductor. *J. Aerospace Sci.* **26**, 249–250.

Chapter III

Akhiezer, A., Liubarskii, G., and Polovin, R. (1959). The stability of shock waves in magnetohydrodynamics. *Soviet Phys. JETP (English Transl.)* **35**, 507–511.

Anderson, E. (1963). "Magnetohydrodynamic Shock Waves." M.I.T. Press, Cambridge, Massachusetts.

Bazer, J. (1958). Resolution of an initial shear flow discontinuity in one-dimensional hydromagnetic flow. *Astrophys. J.* **128**, 686–712.

Bazer, J., and Ericson, W. (1959). Hydromagnetic shocks. *Astrophys. J.* **129**, 758–785.

Cabannes, H. (1960). Attached stationary shock wave in ionized gases. *Rev. Mod. Phys.* **32**, 973–976.

Chu, C., and Lynn, Y. (1963). Steady magnetohydrodynamic flow past a nonconducting wedge. *AIAA J.* **1**, 1062–1067.

Ericson, W., and Bazer, J. (1960). On certain properties of hydromagnetic shocks. *Phys. Fluids* **3**, 631–640.

Levy, T. (1968). Écoulement magnétohydrodynamique stationaire autour d'un dièdre non conducteur dans le cas de champs croisés. *J. Mécanique* **7**, 583-597.

Lust, R. (1953). Magnetohydrodynamische Stoszwellen in einem Plasma Unendlicher Leitfahigkeit. *Z. Naturforsch.* **8**, 277–284.
Lust, R. (1955). Stationare Magnetohydrodynamische Stosswellen Beliebiger Starke. *Z. Naturforsch.* **10**, 125.
Lynn, Y. (1965). Magnetohydrodynamic shock relations in nonaligned flows. Rept. NYO-1480-29, MF-47. New York Univ., New York.
Polovin, R. (1961). Shock waves in magnetohydrodynamics. *Soviet Phys. Usp. (English Transl.)* **3**, 677–688.
Syrovatskii, S. (1959). The stability of shock waves in magnetodynamics. *Soviet Phys. JETP (English Transl.)* **35**, 1024–1027.
Thibault, R. (1962). Sur le théorème de Zemplen en magnétodynamique des fluides. *Compt. Rend.* **255**, 834–836.

Chapter IV

Burgers, J. (1957). Perturbation of a shock wave into a magnetic fields in magnetohydrodynamics. In "Magnetohydrodynamics" (R. K. M. Landshoff, ed.). Stanford Univ. Press, Stanford, California.
Fabri, J., and Siestrunck, R. (1960). Contribution à la théorie aérodynamique du débitmètre électromagnétique. *Assoc. Tech. Maritime Aeron.* **60**, 333–348.
Germain, P. (1959). Contribution à la théorie des ondes de choc en magnétodynamique des fluides. ONERA Publ. No. 97.
Gilbarg, D. (1951). The existence and limit behavior of the one-dimensional shock layer. *Am. J. Math.* **73**, 256–274.
Harris, L. (1960). "Hydromagnetic Channel Flows." M.I.T. Press, Cambridge, Massachusetts.
Meyer, R. (1952). On waves of finite amplitude in ducts. *Quart. J. Appl. Mech. Math.* **5**, 257–291.
Naze, J. (1961). Action des champs magnétiques sur l'écoulement d'un fluide conducteur d'électricité dans une tuyère. *Astronaut. Acta* **7**, 261–275.
Rubin, S. (1965). Magnetogasdynamic channel. *AIAA J.* **3**, 277–283.

Chapter V

Bois, P. A. (1970). Influence sur quelques types de champs magnétiques de l'écoulement d'un fluid conducteur autour d'une sphère. *J. Mécanique* **9**, 35–59.
Grad, H. (1960). Reducible problems in magneto-fluid dynamic steady flows. *Rev. Mod. Phys.* **32**, 830–847.
Greenspan, H. (1960). Flat plate drag in magnetohydrodynamic flow. *Phys. Fluids* **3**, 581–587.
Greenspan, H., and Carrier, G. (1959). The magnetohydrodynamic flow past a flat plate. *J. Fluid Mech.* **6**, 77–96.
Hasimoto, H. (1959). Viscous flow of a perfectly conducting fluid with a frozen magnetic field. *Phys. Fluids* **2**, 238.
Hasimoto, H. (1960). Magnetohydrodynamic wakes in a viscous conducting fluid. *Rev. Mod. Phys.* **32**, 860–866.

Imai, I. (1960). On flows of conducting fluids past bodies. *Rev. Mod. Phys.* **32**, 992–999.
Johnson, W. (1966). A boundary value problem from the theory of magnetohydrodynamic flow. *Arkiv Rational Mech. Anal.* **22**, 355–363.
Lagerstrom, P. (1961). Méthodes asymptotiques pour l'étude des équations de Navier-Stokes. Lecture notes (polycopiées). Univ. of Paris, Institut Henri Poincaré, Paris, France.
Levy, T. (1967). Écoulement lent d'un fluide visqueux et conducteur, autour d'une coque sphérique, en présence d'un champ magnétique. *J. Mécanique* **6**, 529–545.
Levy, T. (1970). Perturbation d'un écoulement de Stokes autour d'une sphère creuse en magnétohydrodynamique. *J. Mécanique* **9**, 189–227.
Ludford, G., and Murray, J. (1960). On the flow of a conducting fluid past a magnetised sphère. *J. Fluid Mech.* **7**, 516–528.
Mitchner, M. (1959). Magnetohydrodynamic flow with a transverse magnetic field. *In* "The Magneto-Dynamics of Conducting Fluids" (D. Bershader, ed.), pp. 61–90. Stanford Univ. Press, Stanford, California.
Peyret, R. (1960). Sur une correspondance entre certains écoulements de magnétodynamique des fluides et ceux de la dynamique des gaz. *Compt. Rend.* **250**, 1971–1973.
Peyret, R. (1961). Sur les régimes de transition dans certains écoulements d'un fluide parfaitement conducteur. *Compt. Rend.* **252**, 2816–2818.
Seebass, R., and Tamada, K. (1965). The distortion of a magnetic field by the flow of a conducting fluid past a circular cylinder. *J. Fluid Mech.* **22**, 561-578.
Watson, G. (1944). "A Treatise on the Theory of Bessel Functions." Cambridge Univ. Press, London and New York.
Wilson, D. (1964). Dual solutions of Greenspan-Carrier equations. *J. Fluid Mech.* **18**, 161–166.

Chapter VI

McCune, J., and Resler, E. (1960). Compressibility effects in magneto-aerodynamic flows past thin bodies. *J. Aerospace Sci.* **27**, 493–503.
Sears, W., and Resler, E. (1958). Theory of thin airfoils in fluids of high electrical conductivity. *J. Fluid Mech.* **5**, 257–273.
Stewartson, K. (1960). On the notion of a non-conducting body through a perfectly conducting fluid. *J. Fluid Mech.* **8**, 82–96.

INDEX

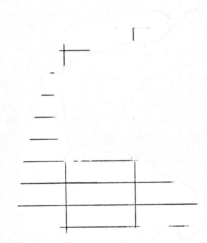